A GUIDE TO NUCLEAR REACTION AND NUCLEAR PARTICLES

A GUIDE TO NUCLEAR REACTION AND NUCLEAR PARTICLES

K.P. Das

CYBER TECH PUBLICATIONS
4264/3, Ansari Road, Daryaganj, New Delhi-110002 (India)
Ph.: 011-23244078, 011-43559448 Fax: 011-23280028
E-Mail: cyberpublicationsdelhi@yahoo.com
Website: www.cybertechpublications.com

A GUIDE TO NUCLEAR REACTION AND NUCLEAR PARTICLES

K.P. Das

First Edition 2011

Published by :

G.S. Rawat for **Cyber Tech Publications**
4264/3, Ansari Road, Daryaganj, New Delhi-110002 (India)
Ph.: 011-23244078, 011-43559448 Fax: 011-23280028
E-Mail: cyberpublicationsdelhi@yahoo.com
Website: www.cybertechpublications.com

PREFACE

This books "A Guide to Nuclear Reaction and Nuclear Particles" has been specially written to the requirements syllabus of the students of all Indian Universities. The subject matter has been discussed in such a simple way that the students will find no difficulty to understand it.

This book contains complete theory and fairly large number of examples. Sufficient problems has also been selected from various universities examination papers. The author shall be thankful to the readers who point out errors and omissions which inspite of all care might have been there.

The proofs of various theorems and examples has been given with minute details. The author will feel highly obliged to all students for their constructive suggestions and healthy criticism of the book which will go a long way in the imporovement of the text.

—Author

CONTENTS

1

Theory of the Nuclear Reactions and Cross-sections

INTRODUCTION

In stripping reaction the captured nucleon, say the neutron, occupies one of the available quantum states in the target nucleus and the other, say the proton, carries some information concerning this state. It was pointed out first by Bethe and Butler in 1952 that stripping reactions in some cases could be used to measure the purity of shell model states. Conversely in pick-up reaction the picked up nucleon is removed from an occupied quantum state and the emerging deuteron conveys some information about this level. Thus we see that the stripping can be used the study unoccupied nuclear states and pick-up to investigate occupied states.

For example let us consider the reaction ${}_6C^{12}$ (d, p) ${}_6C^{13}$. According to the shell model C^{12} has spin 0 and positive parity and C^{13} has spin 1/2 and negative parity. By the use of parity conservation law

$$(-1) = +(1)(-1)^{l_c}.$$

Thus l_c of the captured particle must be odd.

We know that the angular momentum of the captured particle

$$I_c = I_f - I_i = \frac{1}{2} - 0 = \frac{1}{2}.$$

Since $\quad I_c = I_c + s_c \quad$ or $\quad I_c = I_c - s_c$

$$\therefore \quad I_c = \frac{\vec{1}}{2} - \frac{\vec{1}}{2} = 0 \text{ or } 1.$$

The $l_c = 0$ is not acceptable as it violates parity law, hence $l_c = 1$. Thus if the shell model prediction of C^{13} spin and parity is correct, then we should observe a peak around 17° in the differential scattering cross-section corresponding to $l = 1$ capture. This is exactly what is observed experimentally.

In the reaction P^{31} (d, p) P^{32}, the target nucleus has angular momentum $I_i = \frac{1}{2}^+$ (an odd $2s_{1/2}$ proton) and the ground state of P^{32} is expected to be formed by adding a neutron in a $d_{3/2}$ state. The two l-values can combine to $I_f = I_p + I_n = 1^+$ or 2^+. If we apply the model independent selection rule we find that the ground state (1^+ state) can be reached both by an $l = 0$ and $l = 2$ neutron, whereas the shell model state $d_{3/2}$ requires that $l = 2$. Bethe and Butler raised the question : Will a stripping experiment yield an angular distribution consistent with $l = 2$, with $l = 0$ or with a mixture of these?

In the same year, after performing experiment, Parkinson, Beach and King found an angular distribution consistent with $l = 2$ and very little (if any) trace of $l = 0$ contribution. In this particular example the test is very sensitive because $l = 0$ stripping gives a cross-section that is one order of magnitude larger at the maximum than does $l = 2$ stripping under similar circumstances. The result is, therefore, a great tribute to the shell model.

KINEMATICS OF THE NUCLEAR REACTION

There are certain restrictions in the conservation of energy and momentum. These restrictions are called kinematic restrictions and this mathematical method is known as *kinematics*. Consider a nuclear reaction

$$X + x \rightarrow Y + y, \qquad ...(1)$$

where X, x, Y and y are the target nucleus, bombarding particle, product nucleus and product particle, respectively. It will be assumed that target nucleus is at rest so it has no kinetic energy. Since total energy is conserved in a nuclear reaction, therefore, we get

$$Mxc^2 + (E_x + m_x c^2) = (E_Y + M_Y c^2) + (E_y + m_y c^2 \qquad ...(2)$$

where m_x, m_y, M_Y. My all represent respective masses of incident particle, target nucleus, product particle and product nucleus. We now introduce a quantity Q where

$$Q = E_Y + E_y - E_x. \qquad ...(3)$$

From eqns (2) and (3), we have

$$(M_x + m_z - M_Y - m_y)c^2 = Q. \qquad ...(4)$$

The quantity Q is called the energy balance of the reaction or more commonly *Q-value of the reaction.* If this is +ve the reaction is said to be *exoergic.* This occurs if sum of the masses of incident particle and target nucleus is greater than that of masses of the product nuclei. The K.E. of the product nuclei being greater than that of the incident particle. If Q is – ve the reaction is said to be *endoergic, i.e.,* energy must be supplied usually as a K. E. of the incident particle.

The term E_Y in eqn. (2) represents the recoil energy of the product nucleus. It is usually small and hard to measure but can be eliminated by taking into account the conservation of momentum. In an experiment to measure a Q value, the bombarding energy E_x and the energy of the ejected particle E_y at some specified angle θ are measured. Thus by applying the laws of conservation of momentum, we have

$$m_{xyx} = M_Y V_Y \cos\phi + m_y y_y \cos\theta \qquad ...(5)$$

$$\text{and } M_Y V_Y \sin\phi = m_y y_y \sin\theta, \qquad ...(6)$$

where v_x, v_y and V_Y are the velocities of incident particle, ejected particle and of product nucleus respectively.

From eqns (5) and (6) eliminating ϕ, we have

$$M_Y{}^2 V_Y{}^2 = mM_Y^2 Y_Y^2 = m_x^2 + m_y^2 y_y^2 - 2m_x m_y v_x v_y \cos\theta. \qquad ...(7)$$

$$\text{Since } E_x = \frac{1}{2} m_x v_x^2,\ E_y = \frac{1}{2} m_y v_y^2 \text{ and } E_Y = \frac{1}{2} m_Y V_Y^2$$

Hence after eliminating v_x, v_y and V_Y, we get

$$2E_Y M_Y = 2E_x m_x + 2E_y m_y - 4(m_x m_y E_x E_y)^{1/2} \cos\theta$$

$$\text{or } E_Y = E_x \cdot \frac{m_x}{M_Y} + E_y \frac{m_y}{M_Y} - \frac{2}{M_Y}\left(m_x m_y E_x E_y\right)^{1/2} \cos\theta \qquad ...(8)$$

Substituting the value of E_Y in equation (3), we get

$$Q = E_x \frac{m_x - M_Y}{M_Y} + E_y \frac{m_y + M_Y}{M_Y} - \frac{2}{M_y}\left(m_x m_y E_x E_y\right)^{1/2} \cos\theta \qquad ...(9)$$

This is known as Q-value reaction. It gives the desired relation between the energy released and the measured quantities E_x, E_y and θ. For a special case when we are observing the outcoming particle y at 90° to a collimated beam of projectile, the above relation reduces to

$$Q = E_y\left(\frac{1 + m_y}{M_Y}\right) - E_x\left(\frac{1 - m_x}{M_Y}\right). \qquad ...(10)$$

With the modern techniques, it is possible to measure Q values, for nuclear reaction producing charged particles, to an accuracy of 1 part in a thousand or better.

General Solution of the Q-equation

It is clear from eqn. (9) that the variation of E_y with E_x, for a fixed Q, can also be shown by regarding the Q equation as a quadratic in $\sqrt{E_y}$. Thus its solution can be put in the form

$$\sqrt{E_y} = u \pm \sqrt{\left(u^2 + v\right)}, \quad \text{...(11)}$$

where $u = \frac{\sqrt{\left(m_x m_y E_x\right)}}{M_Y + m_y} \cos\theta$ and $v = \frac{M_Y Q + E_x\left(M_Y - m_x\right)}{M_Y + m_y}$.

The energetically possible reactions are those for which $\sqrt{E_y}$ is real and positive. Nuclear reactions are divided energetically as follows

1. Exoergic Reactions

These reactions are possible even for $E_x = 0$. Thus for $E_x \rightarrow 0$, equation (11) gives

$$E_y = Q\frac{M_Y}{\left(m_y + M_Y\right)} (Q > 0). \quad \text{...(12)}$$

The kinetic energy E_y is the same for all angles θ. When $E_x \rightarrow 0$.

$$Q = E_y + E_Y \text{ and } \theta + \phi = 180°.$$

Commonly projectile x is lighter than the product nucleus Y. Thus ν is +ve for all values of the bombarding energy and only one of the two solutions of eqn., (11) will be positive. Thus E_y to be single valued $Q > 0$ and $M_y > m_x$. E_y depends on cos θ and is smallest in the backward direction θ = 180°.

2. Endoergic Reactions

All endoergic reactions have negative Q-values. When $E_x \rightarrow 0$, eqn. (11) gives $u^2 + v = -ve$ and hence $\sqrt{E_y}$ is imaginary. It means that these reactions are not possible. The smallest value of E_x at which reactions can take place is called the *threshold energy*. The reaction first becomes possible when E_x is large enough to make $u^2 + v = 0$, hence

$$E_x = -Q\left[\frac{m_y + M_Y}{m_y + M_Y - m_x - \left(\frac{m_x m_y}{M_Y}\right)\sin^2\theta}\right].$$

At $\theta = 0$, E_x is minimum and is the threshold energy.

$$\therefore \qquad (E_x)_{th} = -Q\frac{(M_Y + m_y)}{(M_Y + m_y - m_x)} \qquad ...(13)$$

By using relation (4) and an excellent approximation $M_x \geq \frac{Q}{c^2}$, we have

$$(E_x)_{th} = -Q\frac{(m_x + M_X)}{M_X} \qquad ...(14)$$

Thus we see that at the threshold or the reaction particles first appear in the $\theta = 0$ direction with the K.E.

$$E_y = u^2 = (E_x)_{th}\left(\frac{m_x m_y}{m_y + M_Y}\right)^2. \qquad ...(15)$$

As the bombarding energy is raised, particle y begins to appear at $\theta > 0$.

TYPES OF NUCLEAR REACTIONS

When a target nucleus is bombarded by means of energetic projectile particles. The bombardment process may give rise to the following events single or jointly.

(a) Elastic Scattering

The incident particle strikes the target nucleus and leaves without energy loss but, in general, with altered direction of motion. Scattering of α-particles in gold is a good example of this process.

(i) $e^- + {}_6C^{12} \rightarrow {}_6C^{12} + e^-$

(ii) ${}_2He^4 + {}_{79}Au^{197} \rightarrow {}_{79}Au^{197} + {}_2He^4$

(iii) ${}_0n^1 + {}_{92}U^{238} \rightarrow {}_{92}U^{238} + {}_0n^1$

(b) Inelastic Scattering

The scattered particle may loss K E in excess of that required for an elastic collision with the nucleus. This being corresponding increase in the internal energy of the nucleus which is excited to a higher quantum state. This inelastic scattering can be represented by the well known example.

(i) ${}_0n^1 + {}_{82}Pb^{208} \rightarrow {}_{82}Pb^{208*} + {}_0n^1$

(ii) ${}_3Li^7 + {}_1H^1 \rightarrow ({}_3Li^7)^* + {}_1H^1$

(iii) ${}_1H^1 + {}_7N^{14} + 7N^{14*} + {}_1H^1.7N^{14*} \rightarrow 7N^{14} + y.$

The star (*) is used to indicate that, after scattering, nucleus is left in an excited state. In this present example the excess energy is latter radiated away in the from of a γ-quantum.

(c) Radiative Capture

A particle may combine with a nucleus to produce a new nucleus or a compound nucleus which is in an excited state. The excess energy is emitted in the form of γ-ray photons. This type of process is known as radiative capture example

$${}_{12}Mg^{26} + {}_1H^1 \rightarrow ({}_{13}Al^{27})^* \rightarrow {}_{13}Al^{27} + \gamma.$$

(d) Disintegration

On striking the target nucleus the incident particle is absorbed and a different particle is ejected. The product nucleus differs from target nucleus. The incident particle may be an α-particle, proton, neutron, etc. The product particle may be a charged particle or a neutron. Thus process is known disintegration as

$${}_7N^{14} + {}_2He^4 \rightarrow {}_3O^{17} + {}_1H^1.$$

(e) Photo Disintegration

The γ-rays are absorbed by the target nucleus, exciting it to a higher quantum state. If the energy is high enough, one or more particles may be liberated. An example is

$${}_1H^2 + g \rightarrow {}_1H^1 + {}_0n^1.$$

(f) High Energy Reactions

In the energy range about 150 MeV, spallation process merges into new kinds of reactions in which new kinds of particles (mesons, strange particles) are produced along with neutrons and protons.

(g) Direct Reactions

A collision of an incident particle with the nucleus may immediately pull one of the nucleons out of the target nucleus by the so called *pick-up reaction.* In the inverse process, a bombarding particle composed of more than one nucleon may loss one of them to the target by the so-called *stripping reaction, e.g.*

$$_{29}Cu^{63} + {}_1d^2 \rightarrow {}_{29}Cu^{64} + {}_1H^1.$$

(h) Heavy Ion Reactions

Nuclear reactions induced by heavy ions exhibit the characteristics both of compound nucleus and of the stripping and pick up mechanisms. Example

$$N^{14} + Pb^{207} \rightarrow N^{13} + Pb^{203}. \qquad ...(16)$$

(i) Spontaneous Decay

$\beta-$ and $\alpha-$ decay processes may be regarded as this type of nuclear reactions. In these reactions the total energy of the system is not under the experimenter's control.

(j) Spallation Reactions

On capture a incident particle, a heavy nucleus has sufficient energy for the ejection of several particles. Such a reaction is known as spallation reaction, *e.g.,* the nuclear fission, in which heavy nucleus splits mainly in two nuclei,

Example : $_{92}U^{235} + {}_0n^1 \rightarrow {}_{40}Zr^{98} + {}_{52}Te^{136} + 2{}_0n^1.$

NUCLEAR CROSS SECTION

It is usually expressed in terms of an effective area presented by a nucleus towards the beam of bombarding particles, such that the number of incident particles that would strike such an area, calculated upon a purely geometrical basis, is the number observed to lead to the reaction in question. This effective area is called the cross-section for that reaction. Thus the probability of occurrence of a particular nuclear reaction is described by the effective cross-section for that process. The cross-section may also be denned as :

1. The probability that an event may occur when a single nucleus is exposed to a beam of particles of total flux one particle per unit area.
2. The probability that an event may occur when a single particle is shot perpendicularly at a target consisting of one particle per unit area.

Partial Cross-section

The total nuclear cross-section is the effective area possessed by a nucleus for removing the incident particles from a collimated beam by all possible processes. This can be written as a sum of several partial cross-sections which represent the contributions of the various distinct, independent processes which can remove particles from the incident beam. Thus we have

$$\sigma_t = \sigma_s + \sigma_r, \quad ...(17)$$

where σ_l is the total cross-section, as the cross-section that produces measurable scattering and σ_r, the reaction cross-section. Scattering cross-section can be classified as : inelastic scattering and elastic scattering processes, Thus, we get :

$$\sigma_s = \sigma_{cl} + \sigma_{incl.} \quad ...(18)$$

These partial cross-sections can still be sub-divided. In the case of elastic cross-section separate partial cross-sections cannot be written because of the possibility of interference between them. On the other hand all inelastic scattering processes are incoherent and their cross-sections are additive,

$$\sigma_{incl.} = \sigma_s + \sigma_a + \sigma_n + ... \quad ...(19)$$

Similarly or the partial cross-section for capture may be sub-divided into component parts associated with the various possible nuclear reactions. Hence we can write

$$\sigma_r = \sigma' + \sigma'' + \sigma''' + ... \quad ...(20)$$

Differential Cross-section

The cross-section which defines a distribution of emitted particles with respect to the solid angle is called differential cross-section. It is defined as $d\sigma/d\Omega$ and the partial cross-section for a given process is given-by as follows

$$s = \int \frac{d\sigma}{d\Omega} d\Omega. \qquad ...(21)$$

Determination of Cross-section

Experimental values of a can be obtained by the measurement of the attenuation of an incident beam in a known thickness of material. If a monoenergetic beam consisting of *l* particles per unit time distributed uniformly over an area A is allowed to pass through a thin sheet of target material of thickness Δx and having n nuclei per unit volume then we have the available nuclear target area which is the product of the cross-section per target nucleus (σ) and the number of target nuclei lying with in the area A.

$\therefore$ Probability for one incoming particle to hit a nuclear target = Nuclear target area $\sigma nA\ \Delta x$/Total target area A. If the nuclear reaction produces N light product particles per unit time, then the reaction probability is defined as the ratio N/I. Hence we have

$$\frac{N}{I} = \frac{\sigma nA\ \Delta x}{A} \text{ or } \sigma = \frac{N}{\left(\frac{I}{A}\right)(nA\Delta x)}. \qquad ...(22)$$

Hence the nuclear cross section can be defined as the number of light product particles per unit time per unit incident flux and per target nucleus. If a bombarding particle is producing a variety of light reaction products N_1, N_2,... per unit time, then the total cross-section is defined as

$$\sigma_{total} = \frac{N_1 + N_2 + ...}{\left(\frac{I}{A}\right)(nA\ \Delta x)\cdots} \qquad ...(23)$$

If the incident flux (beam intensity) is I_0 and the flux after penetrating a distance x is I, it will become I – dI after a further penetration of distance dx. This reduction in flux is caused by the various collision events occurred with in this distance dx.

$$\therefore \qquad -dI = N = n\sigma\ dxI \text{ or } -\frac{dI}{I} = \sigma n\ dx. \qquad ...(24)$$

This expression contains the product of σ, the cross-section per nucleus or the *microscopic cross-section* and n, the number of nuclei per unit volume. The product is called the *macroscopic cross-section* and is denoted by Σ. Thus we have

$$-\frac{dI}{I} = \Sigma dx \quad \text{or} \quad I = I_0 e^{-\Sigma_x} \qquad ...(25)$$

The experimental arrangement consists of a source and a detector between which can be placed a slab of the experimental material. If I, is the particle intensity reaching the detector in the absence of the absorbing material and I is the value when the slab of thickness x, containing n atoms per unit volume is present, the total cross-section o can be calculated. In many nuclear reactions the light product particles are not produced in an isotropic manner. We thus define differential cross-section $d\sigma/d\Omega$ in terms of the number of light products dN emitted per unit time in a small solid angle $d\Omega$ at some angle θ with the direction of incident beam. From eqn. (22), we have

$$\frac{1}{I} = \frac{dN}{d\Omega} = \frac{nA\Delta x\left(\frac{d\sigma}{d\Omega}\right)}{A} \quad \text{or} \quad \frac{d\sigma}{d\Omega} = \frac{\frac{dN}{d\Omega}}{\left(\frac{1}{A}\right)(nA\Delta x)}.$$

Macroscopic Cross-section and Mean Free Path

The equation (25) shows that Σ corresponds to linear absorption coefficient and has the dimensions of reciprocal length. The penetration distance λ can be defined as the path length over which the neutrons can travel without suffering a collision with a target. This distance is known as *mean free path.* Since beam intensity is proportional to the beam particle density q, hence we can write eqn. (25) as

$$q = q_0 e^{-\Sigma_x}.$$

$$\text{The average path length } \lambda = \frac{\text{Total path length}}{\text{Total no. of particles}} = \frac{\int_0^\infty x dq}{q_0}$$

$$= \frac{1}{\Sigma}\int_0^\infty (-\Sigma x)e^{-\Sigma_x} d(\Sigma x) = \frac{1}{\Sigma} \qquad ...(26)$$

Thus the average path length λ between successive collisions is equal to the reciprocal of the macroscopic cross-section Σ.

CLASSICAL ANALYSIS OF CROSS-SECTION

If the bombarding particle has an impact parameter (the distance of closest approach) b with respect to the nucleus. The angular momentum of the system will be normal to the relative momentum p of the two particles and of magnitude

$$L = pb = \frac{\hbar b}{\lambda\!\!\!^{-}}.$$

If b > R, the bombarding particle should not have much effect on the target nucleus as it is outside the range of nuclear forces. Thus we expect the most important nuclear interactions to occur with those bombarding particles which have angular momentum less than or equal to the maximum value $\frac{\hbar R}{\lambda\!\!\!^{-}}$. Quantum mechanics requires that the angular momentum is quantized and can adopt the following discrete values $\hbar l$ with $l = 0, 1, 2...$ If we now consider the associated wave $\lambda\!\!\!^{-}$ we see that it may be resolved into partial waves, each corresponding to a value of l.

Blatt and Weisskopf have provided the significance of the angular momenta in a plane wave by dividing up the incident beam into cylindrical zones, such that each zone is associated with a different impact parameter. In the central zone there will only be the particles with an impact parameter b less than $\lambda\!\!\!^{-} = \left(= \frac{\hbar}{p}\right)$. Actually it is not possible to visualize an impact parameter less than the associated wavelength (*Uncertainly principle*). The angular momentum is zero throughout this zone. The next zone contains all particles with impact parameters between $\lambda\!\!\!^{-}$ and $2\lambda\!\!\!^{-}$. Similarly the lth zone contains particles with impact parameters between $l\lambda\!\!\!^{-}$ and $(l + 1)\,\lambda\!\!\!^{-}$. The change in angular momentum (*i.e.*, momentum × impact parameter) on passing from zone l to $l + 1$ is $\hbar l$ to $h(l + 1)$.

The area of the lth zone is obtained from the difference between the areas due to radii $(l + 1)\lambda\!\!\!^{-}$ and $l\lambda\!\!\!^{-}$. This is the cross-sectional area σ_r^l of the lth zone. If it is assumed that each particle hitting the nucleus causes a reaction then the reaction cross-section may be obtained by summing σ_r^l, over all values of l from 0 to l_m

$$\therefore \qquad \sigma_r = \sum_0^{l_m} \pi\lambda\!\!\!^{-2}(2l+1) = \pi\lambda\!\!\!^{-2}\left(l_m+1\right)^2. \qquad ...(27)$$

The maximum value of l, corresponds to maximum impact parameter R, is given by $l_m = R/\lambda\!\!\!^{-}$. Substituting this value of l_m in the above relation we have maximum reaction cross-section

$$\sigma_r = \pi\,(R + \lambda\!\!\!^{-})^2. \qquad ...(28)$$

This result suggests the possibility of nuclear reaction cross-sections which are several orders of magnitude larger than the geometrical cross-section of the nucleus. However, it will be recalled that the incident wave may also be partially reflected and that a transmission factor Tl, depending on the value of l, must be taken into account. The quantum mechanical treatment of the problem gives

$$\sigma_r = \pi \lambda\!\!\!^{-2} \sum_{l=0}^{\infty} (2l+1) T_l . \qquad ...(29)$$

An interesting case is that of a nucleus which has a radius R very much greater than $\lambda\!\!\!^{-}$, *i.e.,* when the high energy particles strike the nucleus. The nucleus is considered as completely absorbent (black) which corresponds to $T_l = 1$. It is for all values of l upto and including $l_m (= R/\lambda\!\!\!^{-})$. In these conditions the cross-section obtained is same as shown by eqn. (28). This is close to the geometrical cross-section when $\lambda\!\!\!^{-}$ is small is comparison to R. At very high energies, the nucleus exhibits a certain transparency to particles which may penetrate it without interaction. For the weaker energies, $\lambda\!\!\!^{-}$ is too large compared with R for the simplifying hypothesis $T_l = 0$ for large l to hold, since even for $l = 0$, $l > R$.

The elastic cross-section depends upon the condition under which the incident wave is reflected. The interference with the incident wave gives rise to amplitudes four times that of the penetrating wave.

$$\sigma_{el}(\max) = 4\pi(R + \lambda\!\!\!^{-})^2. \qquad ...(30)$$

In contrast, when $T_l = 1$ for all the values of l, in the case of the black nucleus previously examined, the minimum value is given as

$$\sigma_{el}(\min) = \pi(R + \lambda\!\!\!^{-})^2 = \sigma_{inel}. \qquad ...(31)$$

The effective total cross-section is, therefore, equal to

$$\sigma_l = \sigma_{el} + \sigma_{inel} = 2\pi (R + \lambda\!\!\!^{-})^2. \qquad ...(32)$$

Thus we see that for black nucleus the total cross-section is twice the geometrical cross-section. The opaque disc of the nucleus projects a shadow and diffraction occurs as in wave optics. If the nucleus is replaced by a disc of radius (R + $\lambda\!\!\!^{-}$) this shadow may be shown to exist. The disc then emits particle waves of the same intensity in the direction of the incident beam but opposite in phase to those of the beam. In the derivation of above theory we have not considered the effect of Coulombs repulsion. The Coulomb repulsion will bring the relative K.E. of the charged particle from T (when particle is far apart from nucleus) to T

– B, when the particle is just touching the nucleus of radius R. If z and Z are the atomic numbers of incident particle and target nucleus respectively, then $B = zZe^2/4\pi\epsilon_0 R$. Thus the Coulomb repulsion slows down the particle so that the initial moment $(2MT)^{1/2}$ is decreased to $[2M(T - B)]^{1/2}$, where M is the reduced mass of the system. Hence the maximum angular momentum.

$$L_m = R[2M(T - B)]^{1/2} = R(2MT)^{1/2}(1 - B/T)^{1/2}$$

$$\therefore \quad b_m = \frac{L_m}{p} = \frac{L_m}{(2MT)^{1/2}} = R\left(1 - \frac{B}{T}\right)^{1/2}.$$

Thus the Coulomb repulsion diminishes the l_m by a factor of $\left(1 - \frac{B}{T}\right)^{1/2}$. The upper limit for the capture of charged particles can be estimated as the area of the disc of radius b_m.

$$\sigma_r = \pi R^2\left(1 - \frac{B}{T}\right). \qquad ...(33)$$

PARTIAL-WAVE ANALYSIS OF REACTION CROSS-SECTION

Consider an incident plane wave of neutrons travelling parallel to the z-axis. This plane wave has a spatial part e^{ikz}, where $k = 2\pi/\lambda$, and can be expressed as a super-position of spherical waves with different orbital angular momentum quantum numbers relative to the target nucleus. The expansion is

$$e^{ikz} = e^{ikr\cos\theta} = \sum_{i=0}^{\infty} i^i \sqrt{[4\pi(2l+1)]}\, j_l(kr)\, Y_{l,0}(\theta) \qquad ...(34)$$

This solution involves Spherical Bessel functions $j_l(kr)$ and spherical harmonics $Y_{l,m}(\theta)$. As there is cylindrical symmetry about the z-axis the order m of the spherical harmonics is zero. If r, the distance of the incident particle from the target nucleus, is very large so that $kr \geq l$, then the form of the function $j_l(kr)$ is

$$j_i(kr) = \frac{\sin\left(kr - \frac{1}{2}l\pi\right)}{kr} = i\,\frac{e^{-i(kr - l\pi/2)} - e^{+(kr - l\pi/2)}}{2kr}.$$

$$\therefore \quad e^{ikz} = \frac{\sqrt{\pi}}{kr}\sum_{l=0} \sqrt{(2l+1)}\, i^{l+1}\left[e^{-i(k_r - l\pi)} - e^{+i(k_r - l\pi/2)}\right] Y_{l,0}(\theta)$$

$$...(35)$$

The first exponential term represents a series of spherical waves coming towards the origin with momentum $k\hbar$ and angular momentum $l\hbar$, the second exponential term represents a series of a outgoing spherical waves with just the same properties. The above expression describes an undisturbed plane wave. If however, a nucleus is located at the origin, the amplitudes and phases of the outgoing spherical waves from the origin are, in general, changed. Let the outgoing part of the partial wave with orbital angular momentum l be changed by a factor η_l. This may be a complex number. This is related to the phase shift by $\eta_l = |\eta_l|\, e^{2i\sigma l}$. Elastic scattering will take place, when $|\eta_l| = 1$. If $|\eta_l| < 1$ both elastic scattering and nuclear reaction will take place. Thus the wave function for the disturbed wave is

$$\phi(r) = \frac{\sqrt{\pi}}{kr} \sum_{l=0}^{\infty} \sqrt{(2l+1)i^{l+1}\left[e^{-i\left(k_{r-l\pi/2}\right)} - \eta_l e^{+i\left(k_{r-l\pi/2}\right)}\right]}\, Y_{l,0}(\theta) \quad ...(36)$$

The elastically scattered wave function ψ_{sc}, which is the difference between the disturbed wave and the incident wave, is given by

$$\psi_{sc} = \psi(r) - e^{ikx}$$

$$= \frac{\sqrt{\pi}}{kr} \sum_{l=0}^{\infty} \sqrt{(2l+1)i^{l+1}(1-\eta_l)e^{i(kr-l\pi/2)}}\, Y_{l,0}(\theta) \quad ...(37)$$

Since $\psi(r) \propto e^{ikz} + f(\theta)\, e^{ikr}.r$, hence

$$f(\theta) = \frac{\sqrt{\pi}}{k} \sum_{l=0}^{\infty} \sqrt{(2l+1)i^{l+1}(1-h_l)e^{-il\pi/2}}\, Y_{l,0}(\theta).$$

The scattered cross-section is obtained by dividing the number N_{sc}. of scattered particles per second by the number of incident particles per unit area per second which for the plan wave e^{ikx} is v, the velocity if the particles. Consider a sphere of radius r_0. If r_0 is very much larger than the range of the nuclear forces, than the number of scattered particles per second N_{sc} is equal to the flux of ψ_{sc} this sphere.

$$\therefore\ N_{sc} = \frac{\hbar}{2iM} \int \left(\frac{\partial \psi_{sc}}{\partial r} \psi_{sc}^{*} - \frac{\partial \psi_{sc}^{*}}{\partial r} \psi_{sc} \right)_{r=r_0} r_0^2 \sin\theta\, d\theta\, d\phi, \quad ...(38)$$

where M is the mass of the incident particle. Here integral is extended over the surface of the sphere. Substituting values of ψ_{sc} and ψ^{*}_{sc},

$$N_{sc} = \frac{\hbar}{2iM}\int\left[f^*(\theta)\frac{e^{-ikr0}}{r_0}f(\theta)\left(ik\frac{ikr_0}{r_0}\frac{1}{r_0}\frac{e^{ikr_0}}{r_0}\right)\right.$$

$$\left.-(\theta)\frac{e^{ikr^0}}{r_0}f^*(\theta)\left(-ik\frac{e^{ikr_0}}{r_0}\frac{1}{r_0}\frac{e^{ikr_0}}{r_0}\right)\right]r_0^2 \sin\theta\, d\theta\, d\phi$$

$$= \frac{h2ik}{2iM}\int f^*(\theta)\, f(\theta).\frac{1}{r_0^2}r_0^2 \sin\theta\, d\theta\, d\phi$$

$$\because \quad \frac{\hbar k}{M} = \left(\frac{h}{2\pi}\right)\left(\frac{2\pi}{\lambda}\right)\left(\frac{1}{M}\right) = \frac{p}{M} = v,$$

$$\therefore \quad N_{sc} = \frac{\hbar k}{M}\int|f(\theta)|^2 \sin\theta\, d\theta\, d\phi = v\int|f(\theta)|^2\, d\Omega \qquad \text{...(39)}$$

Because of the orthogonality and normalization of the $Y_{l,\,0,}$ we get

$$N_{sc} = v\frac{\pi}{k^2}\sum_{i=0}^{\infty}(2l+1)\,|1-\eta_l|^2 \qquad \text{...(40)}$$

Thus the scattering cross-section is given by

$$\sigma_{sc} = \frac{N_{sc}}{v} = \frac{\pi}{k^2}\sum_{i=0}^{\infty}(2l+1)\,|1-\eta_l|^2 \qquad \text{...(41)}$$

If we define

$$\sigma_{l,\,sc} = \pi\lambda\!\!\!^{-2}\,(2l+1)\,|1-\eta_l)^2, \qquad \text{...(42)}$$

then eqn. (41) can be rewritten as

$$\sigma_{sc} = \sum_{i=0}^{\infty} s_{l,\,sc}\,. \qquad \text{...(43)}$$

Thus we see that in the integral cross-sections the contributions of the partial waves for different l are simply added.

We now derive an expression for the reaction cross-section σ_{Re}. It is determined by the number N_{Re}, the number of particles which enter a sphere of radius r_0 around the enter without leaving it again through the entrance channel. It is, therefore, equal to the net flux into this sphere as computed from the complete wave-function $\psi(r)$.

$$N_{Re} = -\frac{h}{2iM}\int\left(\psi^*.\frac{\partial\psi}{\partial r} - \psi\frac{\partial\psi^*}{\partial r}\right)_{r=r0} r_0^2 \sin\theta\, d\theta\, d\phi \qquad \text{...(44)}$$

The minus sign is used. because inward flux N_{Re}, is positive. Using the orthogonality and normalization of the Y_l, 0, we get

$$N_{Re} = v \frac{\pi}{k^2} \sum_{i=0}^{\infty} (2l+1)\left(1-|\eta_l|^2\right) \qquad ...(45)$$

$$\therefore \sigma_{Re} = \frac{\pi}{k^2} \sum_{i=0}^{\infty} (2l+1)\left(1-|\eta_l|^2\right) = \pi ƛ^2 \sum_{i=0}^{\infty} (2l+1)\left(1-|\eta_l|^2\right). \qquad ...(46)$$

Using the definition

$$\sigma_{l, Re} = \pi ƛ^2 \, (2l + 1) \, (1 - |\eta_l|^2). \qquad ...(47)$$

We can represent eqn. (46) as $\sigma_{Re} = \sum_{i=0}^{\infty} \sigma_{l, Re}$. ...(48)

The total cross-section for a given l is the sum of equations (42) and (47)

$$\begin{aligned} \sigma_{l, t} &= \sigma_{l, sc} + \sigma_{l, Re} \\ &= \pi ƛ^2 (2l+1)\left[|1-\eta_l|^2 \_ \left(1-|\eta_l|^2\right)\right] \\ &= 2\pi ƛ^2 \, (2l + 1) \, [1 - R_e(\eta_l)], \end{aligned} \qquad ...(49)$$

where $R_e(\eta_l)$ is the real part of the complex number η_l.

Conclusion

1. If $\eta_l = 0$, then $s_{l, Re}$ takes on its maximum value $\pi ƛ^2(2l+1)$, however the scattering cross-section takes on the finite value. Scattering and reaction cross-sections are then both of the same size.
2. If $\eta_l = -1$, then $\sigma_{l, Re} = 0$ and the scattering cross-section $\sigma_{l, sc}$ is maximum having value $4\pi ƛ^2(2l+1)$.
3. If $h_l = 1$, then $\sigma_{l, sc} = 0$. It also implies that there are no reactions, as there is no case in which reactions occur without elastic scattering.
4. If $|\eta_l| = 1$, then $\sigma_{l, Re} = 0$, but $\sigma_{l, sc}$ may be finite for the complex value of η_l.
5. $|\eta_l|^2$ can never be negative, otherwise the outgoing wave will have a high intensity than the incoming one. Thus the condition

$|\eta_l|^2 \le 1$ insures that the cross-section $\sigma_{l,\,Re}$ does not become negative.

If the incident neutron has a high enough energy, the reduced wavelength $\lambda\!\!\!^{-}$ is small compared to the radius R of the nucleus. Suppose the nucleus is completely black to the incident neutrons, *i.e.*, all neutrons which hit the nucleus are absorbed by it and, therefore, lead to a reaction, all the particles with $l \le R/\lambda\!\!\!^{-}$ strike the nucleus. We can express this as

$$\eta_l = 0 \text{ if } l \le \frac{R}{\lambda\!\!\!^{-}} \text{ and } \eta_l = 1 \text{ if } l > \frac{R}{\lambda\!\!\!^{-}}. \qquad ...(50)$$

Therefore all incident particles with $l \le \frac{R}{\lambda\!\!\!^{-}}$ react and all those with $l > \frac{R}{\lambda\!\!\!^{-}}$ pass by. Thus eqn. (46) gives

$$\sigma_{Re} = \sum_{i=0}^{R/\lambda\!\!\!^{-}} \pi\lambda\!\!\!^{-2}(2l+1) = \pi(R+\lambda\!\!\!^{-})^2. \qquad ...(51)$$

When σ_{Re} has its maximum value, $\sigma_{Re} = \sigma_{Re}$, and thus we obtain the apparently paradoxical result that the total cross-section is twice them geometrical cross-section of the nucleus:

$$\sigma_l = \sigma_{Sc} + \sigma_{Re} = 2\pi(R+\lambda\!\!\!^{-})^2 \sim 2\pi R^2. \qquad ...(52)$$

The physical meaning of the results involves the diffraction of the incident wave to form a shadow behind the absorbing centre. Since any neutron striking the nucleus reacts, therefore, the nucleus will appear opaque to a fast neutron.

COMPOUND NUCLEUS

According to the Bohr, the nuclear reaction takes place in two distinct and independent stages :

(i) *formation of a compound nucleus C and*

(ii) *the disintegration of the compound nucleus into the products of the reaction.* The compound nucleus, which is a many body system of strongly interacting particles, is formed by the amalgamation of an incident particle x with a target nucleus X.

$$X + x \rightarrow C^{*}. \qquad ...(53)$$

The incident particle captured by a nucleus gives up its energy to few nucleons and, as the results of the interaction of these nucleons with all the others, the energy is quickly distributed among all the nucleons of the compound nucleus. The new nucleus thus formed is in excited state. If E_x is the kinetic energy of the incident particle of mass m, the excitation energy of the compound nucleus will be given by

$$E^* = \frac{E_x M}{(m+M)+E_B}, \qquad ...(54)$$

where M is the mass of the target nucleus and E_0 the binding energy of the particle in the compound nucleus. The mode of disintegration of compound nucleus ($C^* \rightarrow Y + y$) is independent of the mode of formation and depends only on its energy, angular momentum and parity. If the different processes lead to the same compound nucleus the decomposition is identical. A compound nucleus, once formed, can decay in a number of different ways, each with its own-intrinsic probability. For example a compound nucleus Al^{27*}, formed by the bombardment of Na^{23} with α-particles or by other methods, can decay at least in following ways : $Al^{27*} \rightarrow (Na^{23} + \alpha)$, $(Mg^{25} + d)$, $(Mg^{26} + p)$, $(Al^{26} + n)$ or $(Al^{27} + \lambda)$. In each case the residual nucleus can usually be left in one of its excited states, leading to still more modes of decay.

The compound nucleus, has a life time which is long ($10^{-14} - 10^{-18}$ sec) compared to the time for a nucleon to traverse a nucleus ($10^{-20} - 10^{-23}$ sec.). This finite life time is because there can always be a statistical fluctuation in the energy distribution which concentrates enough energy on a nucleon to allow to escape.

Energy Levels of Nuclei

Resonances in the yield curve of a nuclear transmutation will occur whenever the energy of the incident particle together with its B.E. in the compound nucleus is equal to and coincides with an energy level of that nucleus. An energy level that corresponds to an excitation energy greater than the B.E. of the particle in the compound nucleus is a *virtual energy level* and the nucleus is than said to be in a *quasi-stationary state*. The most stable energy state of nucleus is the lowest energy level or *ground state* (*zero potential energy*). Corresponds to this state the energy of the nucleus is the B.E. calculated from the mass defect. For excitation energies less than the B.E., the corresponding energy levels are *bound energy levels*. In the normal state of the nucleus, the lowest bound levels are all occupied.

The nucleon will be in one of the normally unoccupied bound states, if the excitation energy is less than about 8 MeV. The region of the virtual levels, contains those energy levels that can be occupied during a nuclear reaction while the nucleus exists in an intermediate state as a compound nucleus. De-excitation of bound energy levels can take place only by emission of γ-rays, whereas de-excitation from a virtual energy level can occur in a variety of ways, either by particle emission or γ-emission. For example, in certain (α, p) reactions discrete energy values are observed amongst the ejected protons, although monoenergetic α-particles are being used in the bombardment of the target.

These experimental results can be understood if one attributes to each energy group a corresponding energy level in the nuclei of their origin. Thus, the largest of the group energies E_0 is associated with the formation of residual nucleus in its lowest or ground state. The smaller energy values (E_1, E_2, E_3) of the groups are associated with residual nuclei which have been left in higher excited energy states. The energy differences between the various groups can be related to the respective energy level spacing in the residual nucleus. This conclusion is supported by the fact that the lower energy groups are accompanied by simultaneous γ-ray emission, whereas the highest energy group is not accompanied by any γ-emission from the residual nucleus.

Level Width and De-excitation

The excited state of the compound nucleus has a definite mean life time before it decays by one of the possible modes of de-excitation. We know that

$$\text{Mean life time } \tau = 1/\text{disintegration constant } \lambda, \qquad \text{...(55)}$$

where λ is connected with the level width Γ by the relation

$$\Gamma = \hbar\lambda \qquad \text{...(56)}$$

$$\therefore \qquad \Gamma = \frac{\hbar}{\tau} \text{ or } \Gamma\,\tau = \hbar. \qquad \text{...(57)}$$

According to the principle of Heisenberg uncertainty

$$\Delta E.\Delta t = \hbar. \qquad \text{...(58)}$$

If we equate the uncertainty in the time measurement Δ t with the mean life time of the excited state, we obtain level width Γ for the uncertainty in the energy of the excited state Δ E. *Hence the level width of the excited state is a spread in energy of the excited level due to the uncertainty in the level.* As the level width is inversely proportional to the mean life time τ, it is clear that a long life time means a very fine and narrow energy level, whereas a short life time means a broad or diffuse energy level of larger level width.

For each individual mode of decay, there is a different probability of decay and, therefore, a different partial width for each decay product. The total width of an energy level is then the sum of the individual partial widths, *i.e.*, $\Gamma = \Gamma_1 + \Gamma_2 + \Gamma_3 + ...$

For bound levels, neglecting internal conversion $\Gamma = \Gamma_\gamma$. But for virtual level $\Gamma = \Gamma_\gamma + \Sigma\Gamma_{particles}$, where the summation includes the widths corresponding to the emission of certain energetically allowed particles. The widths Γ_γ and $\Gamma_{particles}$ are partial widths. For the ground state $\Gamma = 0$.

NUCLEAR TRANSMUTATIONS

In this section reference will be made to specific cases of nuclear reactions with various projectiles.

1. Transmutation by Alpha Particles

(a) (α-p) Reaction: Rutherford first observed in 1919 the transmutation of the nitrogen nucleus by α-particles in which α-particle was captured by a nucleus forming a compound nucleus which disintegrated immediately into a new nucleus by the ejection of a proton. Few of these (α-p) reactions are :

(i) ${}_{21}Sc^{45} + {}_2He^4 \rightarrow ({}_{23}V^{49})^* \rightarrow {}_{22}Ti^{48} + {}_1H^1 - 0.3$ MeV.

(ii) ${}_{19}K^{39} + {}_2He^4 \rightarrow ({}_{21}Sc^{43})^* \rightarrow {}_{20}Ca^{42} + {}_1H^1 - 0.89$ MeV

(iii) ${}_{16}S^{32} + {}_2He^4 \rightarrow ({}_{18}A^{36})^* \rightarrow {}_{17}Cl^{35} + {}_1H^1 - 2.10$ MeV

(iv) ${}_{14}Si^{28} + {}_2He^4 \rightarrow ({}_{16}S^{32})^* \rightarrow {}_{15}P^{31} + {}_1H^1 - 1.92$ MeV

(v) ${}_{13}Al^{27} + {}_2He^4 \rightarrow ({}_{15}P^{31})^* \rightarrow {}_{11}Si^{30} + {}_1H^1 + 2.26$ MeV

(vi) ${}_9F^{19} + {}_2He^4 \rightarrow ({}_{11}Na^{23})^* \rightarrow {}_{10}Ne^{22} + {}_1H^1 + 1.58$ MeV

(vii) ${}_5B^{10} + {}_2He^4 \rightarrow ({}_7N^{14})^* \rightarrow {}_6C^{13} + {}_1H^1 + 4.04$ MeV

Few (α-p) reactions lead to the formation of radiative isotopes that decay by β-emission. For example

$${}_5B^{11} + {}_2He^4 \rightarrow ({}_7N^{15})^* \rightarrow {}_6C^{14} + {}_1H^1 + 0.75 \text{ MeV}$$

followed by ${}_6C^{14} \rightarrow {}_7N^{14} + \beta, - T = 5580$ years.

(b) (α-n) Reaction : The bombardment of beryllium by α-particle with the subsequent emission of neutrons is one of the (α-n) reactions. Few of nuclear reactions known as (α-n) type are :

(i) ${}_{13}Al^{27} + {}_{2}He^{4} \rightarrow ({}_{15}P^{31})^{*} \rightarrow {}_{15}P^{30} + {}_{0}n^{1}$.

(ii) ${}_{9}F^{19} + {}_{2}He^{4} \rightarrow ({}_{11}Na^{23})^{*} \rightarrow {}_{11}Na^{22} + {}_{0}n^{1}$

(iii) ${}_{7}N^{14} + {}_{2}He^{4} \rightarrow ({}_{9}F^{18})^{*} \rightarrow {}_{9}F^{17} + {}_{0}n^{1}$

(iv) ${}_{5}B^{11} + {}_{2}He^{4} \rightarrow ({}_{7}N^{15})^{*} \rightarrow {}_{7}N^{14} + {}_{0}n^{1}$

(v) ${}_{4}Be^{9} + {}_{2}He^{4} \rightarrow ({}_{6}C^{13})^{*} \rightarrow {}_{6}C^{12} + {}_{0}n^{1}$

(vi) ${}_{3}Li^{7} + {}_{2}He^{4} \rightarrow ({}_{5}B^{11})^{*} \rightarrow {}_{5}B^{10} + {}_{0}n^{1}$.

In many of these reactions the product nuclei are left in excited states. Most of the product nuclei formed by (α-n) reaction are unstable isotopes which then disintegrate with the emission of a positron. Few of these are given below with their half lives.

(i) ${}_{15}P^{3} \rightarrow {}_{14}Si^{30} + \beta^{+}$, T = 2.5 min.

(ii) ${}_{14}Si^{28} \rightarrow {}_{13}Al^{27} + \beta^{+}$, T = 4.9 sec.

(iii) ${}_{13}Al^{26} \rightarrow Mg^{26} + \beta^{+}$, T = 7 sec.

(iv) ${}_{11}Na^{22} \rightarrow {}_{10}Ne^{22} + b^{+}$, T = 2.6 years.

(v) ${}_{9}F^{17} \rightarrow {}_{8}O^{17} + \beta^{+}$, T = 1.2 min.

(vi) ${}_{7}N^{13} \rightarrow {}_{6}C^{13} + \beta^{+}$, T = 10.1 min.

(c) When the energy of the a particle is large enough, two or more nucleons may be expelled in a nuclear reaction, processes of the (α, 2n), (α, np), (α, 3n), (α, 4n), (α, 3np), etc., types have been detected.

(d) Radiative Capture : It is also possible, for the compound nucleus formed by the capture of an α-particle to go to a more stable configuration without emitting a particle but with emitting a γ-ray photon.

$${}_{3}Li^{7} + {}_{2}He^{4} \rightarrow ({}_{5}B^{11}) \rightarrow {}_{5}B^{11} + h\nu.$$

2. Transmutation by Protons

(a) (p-α) Reaction : The first artificial transmutation produced by protons was given by Cockcroft and Walton (1932). The nuclear reaction is given by equation

$${}_{3}Li^{7} + {}_{1}H^{1} \rightarrow ({}_{4}Be^{8})^{*} \rightarrow {}_{2}He^{4} + {}_{2}He^{4} + 17.28 \text{ MeV}.$$

Other examples are :

(i) $_{17}Cl^{35} + {}_1H^1 \rightarrow ({}_{18}A^{36})^* \rightarrow {}_{16}S^{32} + {}_2He^4$

(ii) $_{11}Na^{23} + {}_1H^1 \rightarrow ({}_{12}Mg^{24})^* \rightarrow {}_{10}Ne^{20} + {}_2He^4$

(iii) $_9F^{19} + {}_1H^1 \rightarrow ({}_{10}Ne^{20})^* \rightarrow {}_8O^{16} + {}_2He^4$

(iv) $_5B^{11} + {}_1H^1 \rightarrow ({}_6C^{12}) \rightarrow {}_4Be^8 + {}_2He^4$

(v) $_4Be^9 + {}_1H^1 \rightarrow ({}_5B^{10})^* \rightarrow {}_3Li^6 + {}_2He^4$

(vi) $_3Li^6 + {}_1H^1 \rightarrow ({}_4Be^7)^* \rightarrow {}_2He^3 + {}_2He^4$

(b) (p-n) Reaction : The probability of the (p, n) reaction is relatively high if the proton has sufficient energy to penetrate the Coulomb energy barrier. Few examples are :

(i) $_{29}Cu^{35} + {}_1H^1 + \rightarrow ({}_{30}Zn^{66})^* \rightarrow {}_{30}Zn^{65} + {}_0n^1$

(ii) $_{28}Ni^{58} + {}_1H^1 \rightarrow ({}_{29}Cu^{59})^* \rightarrow {}_{29}Cu^{58} + {}_0n^1$

(iii) $_{11}Na^{23} + {}_1H^1 \rightarrow ({}_{12}Mg^{24})^* \rightarrow {}_{12}Mg^{23} + {}_0n^1$

(iv) $_8O^{18} + {}_1H^1 \rightarrow ({}_9F^{19})^* \rightarrow {}_9F^{18} + {}_0n^1$

(v) $_5B^{11} + {}_1H^1 \rightarrow ({}_6C^{12})^* \rightarrow {}_6C^{11} + {}_0n^1$.

In these reactions, the product nucleus is nearly same as the target nucleus but its atomic number is one unit higher. These equations are usually endoergic.

(c) If the incident proton has an energy (> 20 MeV) the compound nucleus has sufficient excitation energy to permit the expulsion to two or more nucleons. Few examples are:

Cu^{63}(p, pn) C^{62}_{11}, Cu^{63} (p, 2d) Zn^{62} and Sr^{88} (p, 3n) Y^{86}.

(d) (p-d) Reaction : It is an another type of disintegration caused by protons Examples of this type of reaction are :

$_3Li^7 + {}_1H^1 \rightarrow {}_3Li^6 + {}_1H^2$

$_4Be^9 + {}_1H^1 \rightarrow {}_4Be^8 + {}_1H^2$.

Reactions of this type are more common with heavier elements and protons of fairly high energy. They appear to be pick-up reaction and do not form any compound nucleus.

(e) Proton Capture : In some cases the compound nucleus is in excited state which returns to its normal state with the gamma ray photon. Examples of this reaction are :

$_3Li^7 + {}_1H^1 \rightarrow ({}_4Be^8)^* \rightarrow {}_4Be^8 + \gamma$

$_6C^{12} + {}_1H^1 \rightarrow ({}_7N^{13})^* \rightarrow {}_7N^{13} + \gamma$

$$_7N^{14} + {}_1H^1 \rightarrow ({}_8O^{15}) \rightarrow {}_8O^{15} + \gamma$$

$$_9F^{19} + {}_1H^1 \rightarrow ({}_{10}Ne^{20})^* \rightarrow {}_{10}Ne^{20} + \gamma$$

$$_{13}Al^{27} + {}_1H^1 \rightarrow ({}_{14}Si^{28})^* \rightarrow {}_{14}Si^{28} + \gamma$$

$$_{24}Cr^{50} + {}_1H^1 \rightarrow ({}_{25}Mn^{51})^* \rightarrow {}_{25}Mn^{51} + \gamma$$

3. Transmutation by Neutrons

Neutrons have no electric charge and can penetrate positively charged nuclei without any experience of repulsive electrostatic forces, hence neutrons have also been found to be extremely effective.

(a) (n-p) Reaction : In this reaction the proton in the nucleus is replaced by a neutron. The mass number is not changed but the charge is decreased by one unit. Few examples of (n-p) reaction produced by slow neutrons are :

$$_7N^{14} + {}_0n^1 \rightarrow ({}_7N^{15})^* \rightarrow {}_6C^{14} + {}_1H^1 + Q$$

followed by $_6C^{14} \rightarrow {}_7N^{14} + \beta^- + E_{max}$.

$$_2He^3 + {}_0n^1 \rightarrow ({}_2He^4)^* \rightarrow {}_1H^3 + {}_1H^1 + Q$$

followed by $_1H^3 \rightarrow {}_2He^3 + \beta^- + E_{max}$.

$$_{17}Cl^{35} + {}_0n^1 \rightarrow ({}_{17}Cl^{36})^* \rightarrow {}_{16}S^{35} + {}_1H^1 + Q$$

followed by $_{16}S^{35} \rightarrow {}_{17}Cl^{35} + \beta^- + E_{max}$

Other (n-p) reactions performed with fast neutrons have Q-values negative. Few examples are :

$$_{13}Al^{27} + {}_0n^1 \rightarrow ({}_{13}Al^{28})^* \rightarrow {}_{12}Mg^{27} + {}_1H^1$$

$$_{16}S^{32} + {}_0n^1 \rightarrow ({}_{16}S^{33})^* \rightarrow {}_{15}P^{32} + {}_1H^1$$

$$_{29}Cu^{65} + {}_0n^1 \rightarrow ({}_{29}Cu^{66})^* \rightarrow {}_{28}Ni^{65} + {}_1H^1$$

$$_{30}Zn^{64} + {}_0n^1 \rightarrow ({}_{30}Zn^{65})^* \rightarrow {}_{29}Cu^{64} + {}_1H^1.$$

In all these reactions the product nuclei are radioactive producing finally nuclei identical with the initial target nuclei.

(b) (n-α) Reaction : In this type of reactions, the capture of slow neutron results in the emission of an α-particle. Few examples are :

(i) $$_7N^{14} + {}_0n^1 \rightarrow ({}_7N^{15}) \begin{cases} {}_5B^{11} + {}_2He^4 \\ {}_3Li^7 + {}_2He^4 + {}_2He^4. \end{cases}$$

(ii) $_{13}Al^{27} + {_0}n^1 \rightarrow ({_{13}}Al^{28})^* \rightarrow {_{11}}Na^{24} + {_2}He^4$.

(iii) $_5B^{10} + {_0}n^1 \rightarrow ({_5}B^{11})^* \rightarrow {_3}Li^7 + {_2}He^4$

(iv) $_3Li^6 + {_0}n^1 \rightarrow ({_3}Li^7)^* \rightarrow {_1}H^3 + {_2}He^4$

The capture of fast neutrons by heavier nuclei results in the emission of an alpha particle, the product nucleus is usually radioactive. Some examples are :

$$_{11}Na^{23} + {_0}n^1 \rightarrow ({_{11}}Na^{24})^* \rightarrow {_9}F^{20} + {_2}He^4$$

followed by $_9F^{20} \rightarrow {_{10}}Ne^{20} + \beta^{-1}$, $T = 10.7$ sec.

$$_{13}Al^{27} + {_0}n^1 \rightarrow ({_{13}}Al^{28})^* \rightarrow {_{11}}Na^{24} + {_2}He^4$$

followed by $_{11}Na^{24} \rightarrow {_{12}}Mg^{24} + \beta^-$, $T = 15.0$ hr.

(c) (n-d) and (n-t) Reactions : (n, d) reactions have been detected as a result of the bombardment of elements by 90 MeV neutrons. These reactions do not produce any compound nucleus and are direct reactions of the pick-up type. (n, t) reactions can take place in a similar manner. Tritium found in atmosphere is most likely being produced by the following reaction :

$$_7N^{14} + {_0}n^1 \rightarrow {_6}C^{12} + {_1}H^3.$$

(d) (n-λ) Reaction : The radioactive capture reaction with slow neutrons is probably the most common nuclear process. It takes place with nearly all elements. The product is an isotope of the target nuclide with mass number larger by unity. Few examples are :

$$_1H^1 + {_0}n^1 \rightarrow ({_1}H^2)^* \rightarrow {_1}H^2 + \gamma$$

$$_1H^2 + {_0}n^1 \rightarrow ({_1}H^3)^* \rightarrow {_1}H^3 + \gamma.$$

The capture of a neutron increases the neutron to proton ratio. If this ratio in the target nucleus is already near the upper limit of stability, the product nucleus is radioactive usually emitting β-particle. Few examples are :

$$_{45}Rh^{103} + {_0}n^1 \rightarrow ({_{45}}Rh^{104})^* \rightarrow {_{45}}Rh^{104} \text{ (42 sec)} + \gamma$$

$$_{49}In^{115} + {_0}n^1 \rightarrow ({_{49}}In^{116})^* \rightarrow {_{49}}In^{116} \text{ (54 min} + \gamma$$

$$_{92}U^{238} + {_0}n^1 \rightarrow ({_{92}}U^{239})^* \rightarrow {_{92}}U^{239} \text{ (23.5 min)} + \gamma$$

$$_{29}Cu^{65} + {_0}n^1 \rightarrow ({_{29}}Cu^{66})^* \rightarrow {_{29}}Cu^{66} \text{ (5.1 min)} + \gamma$$

$$_{79}Au^{197} + {_0}n^1 \rightarrow ({_{79}}Au^{198})^* \rightarrow {_{79}}Au^{198} \text{ (2.7 days)} + \gamma.$$

(e) **Neutron—Three or More Particles :** If the energy of the incident neutron approaches 30 MeV, sufficient energy available to overcome the electrostatic potential barrier, three neutrons or even two neutrons and a proton are ejected from the compound nucleus. When neutrons (~ 100 MeV) are used as projectiles, nuclei of moderate mass number may undergo *spallation* and those of high mass number, *e.g.*, Bi and *Pb suffer fission* probably accompanied by spallation.

(f) **(e-2n) Reaction :** In this type of reaction, one neutron is captured by the nucleus and two neutrons are emitted leaving the charge of the nucleus unchanged and less the mass number by one unit. Thus the residual nucleus is the isotope of the target nucleus. Q-value of the reaction is negative and fast neutrons are needed to bring about this reaction. In most of the cases residual nucleus is unstable and followed by position emission. Few examples are:

$${}_6C^{12} + {}_0n^1 \rightarrow ({}_6C^{13})^* \rightarrow {}_6C^{11}\ (20.4\ \text{min}) + 2\ {}_0n^1$$

$${}_{19}K^{39} + {}_0n^1 \rightarrow ({}_{19}K^{40})^* \rightarrow {}_{19}K^{38}\ (7.5\ \text{min}) + 2\ {}_0n^1$$

$${}_{51}Sb^{121} + {}_0n^1 \rightarrow ({}_{51}Sb^{122})^* \rightarrow {}_{51}Sb^{120}\ (16.6\ \text{min}) + 2\ {}_0n^1.$$

${}_{80}Hg^{197}$, obtained in the ${}_{80}Hg^{198}\ (n, 2n){}_{80}Hg^{197}$ process, decays by equivalent orbital electron capture. The emission of an electron occurs in a number of instances, especially where neutron to proton ratio is at the upper limit of the stability for the given atomic number. Original target nuclei of this type are Ge^{76}, Cd^{116}, Ce^{142} and U^{238}.

4. Transmutation by Deuterons

A great many nuclear reactions have been observed with high energy deuterons as the bombarding particles, because about 2 MeV is sufficient to cause its rupture into a neutron and proton.

(a) **(d-p) Reaction :** Reactions of this type are quite common, and the product is always an isotope of the target element with mass number one unit higher. Few examples are :

$${}_6C^{12} + {}_1H^2 \rightarrow ({}_7N^{14})^* \rightarrow {}_6C^{13} + {}_1H^1$$

$${}_3Li^7 + {}_1H^2 \rightarrow ({}_4Be^9)^* \rightarrow {}_3Li^8 + {}_1H^1\ \text{(electron-decay)}$$

$${}_{11}Na^{23} + {}_1H^2 \rightarrow ({}_{12}Mg^{25})^* \rightarrow {}_{11}Na^{24} + {}_1H^1\ \text{(electron-decay)}$$

$${}_{15}P^{31} + {}_1H^2 \rightarrow ({}_{16}S^{33})^* \rightarrow {}_{15}P^{32} + {}_1H^1$$

$$_{48}Cd^{114} + {}_1H^2 \rightarrow ({}_{49}In^{116})^* \rightarrow {}_{48}Cd^{115} + {}_1H^1$$
$$_{83}Bi^{209} + {}_1H^2 \rightarrow ({}_{84}Po^{211})^* \rightarrow {}_{83}Bi^{210} + {}_1H^1.$$

(b) (d-α) Reaction : Because of the high nuclear potential energy barrier for α-particles, which increases with Z, reactions of this type are observed only with high energy deuterons and elements of fairly low Z. Few examples are:

$$_3Li^6 + {}_1H^2 \rightarrow ({}_4Be^8)^* \rightarrow {}_2He^4 + {}_2He^4$$
$$_8O^{16} + {}_1H^2 \rightarrow ({}_9F^{18})^* \rightarrow {}_7N^{14} + {}_2He^4$$
$$_{10}Ne^{20} + {}_1H^2 \rightarrow ({}_{11}Na^{2q})^* \rightarrow {}_9F^{18} + {}_2He^4$$
$$_{12}Mg^{26} + {}_1H^2 \rightarrow ({}_{13}Al^{28})^* \rightarrow {}_{11}Na^{24} + {}_2He^4$$
$$_{13}Al^{27} + {}_1H^2 \rightarrow ({}_{14}Si^{29})^* \rightarrow {}_{12}Mg^{25} + {}_2He^4.$$

(c) The probability of radiative capture reaction is quite small because the extra nucleons in a compound are in virtual states. When the energy of the incident deuterons is about 20 MeV or more, reactions of the types (d, 2n), (d, 3n), (d, 2p) and (d, pα) are possible.

(d) (d-t) Reaction : Although the probability of the reaction of this type is low, some processes are known. Among these are:

$$_3Li^7 + {}_1H^2 \rightarrow ({}_4Be^9)^* \rightarrow {}_3Li^6 + {}_1H^3$$
$$_{15}P^{31} + {}_1H^2 \rightarrow ({}_{16}S^{83})^* \rightarrow {}_{15}P^{30} + {}_1H^3$$
$$_{47}Ag^{109} + {}_1H^2 \rightarrow ({}_{48}Cd^{111})^* \rightarrow {}_{47}Ag^{108} + {}_1H^3$$
$$_4Be^9 + {}_1H^2 \rightarrow ({}_5B^{11})^* \rightarrow {}_4Be^8 + {}_1H^3.$$

(e) (d-n) Reaction : A large number of transmutations of this

$$_3Li^7 + {}_1H^2 \rightarrow ({}_4Be^9)^* \rightarrow {}_4Be^8 + {}_0n^1$$
$$_4Be^9 + {}_1H^2 \rightarrow ({}_5B^{11})^* \rightarrow {}_5B^{10} + {}_0n^1$$
$$_6C^{12} + {}_1H^2 \rightarrow ({}_7N^{14})^* \rightarrow {}_7N^{13} + {}_0n^1 \quad \text{(position decay)}$$

When two deuterons interact both the (d, n) and the (d, p) reactions have been observed.

$$_1H^2 + {}_1H^2 \rightarrow ({}_2He^4)^* \begin{cases} {}_1H^3 + {}_1H^1 + 4.02 \text{ MeV} \\ {}_2He^3 + {}_0n^1 + 3.25 \text{ MeV} \end{cases}$$

The tritium is unstable and has a half life of about 12 years.

At high energies (> 50 MeV), the (d, n) reaction becomes a stripping process that does not involve compound nucleus formation.

5. Transmutation of Particles of Mass Three

Some reactions obtained with tritium and helium 3 are given below :

$${}_{3}Li^{6} + {}_{1}H^{3} \rightarrow ({}_{4}Be^{9})^{*} \rightarrow {}_{3}Li^{7} + {}_{1}H^{2}$$

$${}_{27}Co^{59} + {}_{1}H^{3} \rightarrow ({}_{28}Ni^{62})^{*} \rightarrow {}_{27}Co^{60} + {}_{1}H^{2}$$

$${}_{29}Cu^{63} + {}_{1}H^{3} \rightarrow ({}_{30}Zn^{66})^{*} \rightarrow {}_{29}Cu^{64} + {}_{1}H^{2}$$

$${}_{1}H^{2} + {}_{1}H^{3} \rightarrow ({}_{2}He^{5})^{*} \rightarrow {}_{2}He^{4} + {}_{0}n^{1}$$

$${}_{16}S^{32} + {}_{1}H^{3} \rightarrow ({}_{17}Cl^{35})^{*} \rightarrow {}_{17}Cl^{34} + {}_{0}n^{1}$$

$${}_{13}Al^{27} + {}_{1}H^{3} \rightarrow ({}_{14}Sl^{30})^{*} \rightarrow {}_{12}Mg^{27} + {}_{2}He^{3}$$

$${}_{3}Li^{7} + {}_{1}H^{3} \rightarrow ({}_{4}Be^{10})^{*} \rightarrow {}_{2}He^{6} + {}_{2}He^{4}$$

$${}_{1}H^{2} + {}_{2}He^{3} \rightarrow ({}_{3}Li^{5})^{*} \rightarrow {}_{2}He^{4} + {}_{1}H^{1}$$

$${}_{1}H^{3} + {}_{2}He^{3} \rightarrow ({}_{3}Li^{6})^{*} \rightarrow {}_{2}He^{5} + {}_{1}H^{1}$$

$${}_{2}He^{3} + {}_{2}He^{3} \rightarrow ({}_{4}Be^{6})^{*} \rightarrow {}_{3}Li^{5} + {}_{1}H^{1}$$

$${}_{7}N^{14} + {}_{2}He^{3} \rightarrow ({}_{9}F^{17})^{*} \rightarrow {}_{9}F^{16} + {}_{0}n^{1}$$

$${}_{6}C^{12} + {}_{2}He^{3} \rightarrow ({}_{8}O^{15})^{*} \rightarrow {}_{7}N^{13} + {}_{1}H^{2}$$

$${}_{6}C^{12} + {}_{2}He^{3} \rightarrow ({}_{8}O^{16})^{*} \rightarrow {}_{7}N^{13} + {}_{1}H^{3}$$

$${}_{7}N^{14} + {}_{2}He^{3} \rightarrow ({}_{9}F^{17})^{*} \rightarrow {}_{7}N^{13} + {}_{2}He^{4}.$$

6. Transmutation by Radiation

Reactions of this type brought about by high energy radiations, have been referred to as photo disintegrations. To produce this process the mean of the incident photon must exceed the binding energy of the particle to be ejected. If the energy of the γ-radiation is not sufficient to remove even neutron, the nucleus may enter one of its excited states. The excess energy is then emitted as radiation. Examples are :

$${}_{1}H^{2} + \gamma \rightarrow ({}_{1}H^{2})^{*} \rightarrow {}_{1}H^{1} + {}_{0}n^{1}$$

$${}_{4}Be^{9} + \gamma \rightarrow ({}_{4}Be^{9})^{*} \rightarrow {}_{4}Be^{8} + {}_{0}n^{1}$$

$${}_{15}P^{31} + \gamma \rightarrow ({}_{15}P^{31})^{*} \rightarrow {}_{15}P^{30} + {}_{0}n^{1}$$

$${}_{4}Be^{9} + \gamma \rightarrow ({}_{4}Be^{9})^{*} \rightarrow {}_{3}Li^{8} + {}_{1}H^{1}$$

$${}_{12}Mg^{25} + \gamma \rightarrow ({}_{12}Mg^{25})^{*} \rightarrow {}_{11}Na^{24} + {}_{1}H^{1}$$

$${}_5B^{10} + \gamma \rightarrow ({}_5B^{10})^* \rightarrow {}_4Be^8 + {}_1H^2$$
$${}_5B^{11} + \gamma \rightarrow ({}_5B^{11})^* \rightarrow {}_4Be^8 + {}_1H^3$$
$${}_6C^{12} + \gamma \rightarrow ({}_6C^{12})^* \rightarrow {}_4Be^8 + {}_2He^4$$
$${}_{36}Kr^{83} + \gamma \rightarrow ({}_{36}Kr^{83})^* \rightarrow {}_{36}Kr^{83} + \gamma'$$
$${}_{38}Sr^{87} + \gamma \rightarrow ({}_{38}Sr87)^* \rightarrow {}_{38}Sr^{87} + \gamma'.$$

One nucleon either neutron or proton can be emitted when γ-rays are of energy of the order of 10 MeV. At high energies reactions of the types (γ, np), (γ, 2n), (γ, n 2p) and others, involving the ejection of two or three or more particles, have been observed. The examples are :

$${}_8O^{16} + \gamma \rightarrow ({}_8O^{16})^* \rightarrow {}_6C^{11} + {}_0n^1 + {}_0n^1 + {}_0n^1 + {}_1H^1 + {}_1H^1$$
$${}_6C^{12} + \gamma \rightarrow ({}_6C^{12})^* \rightarrow {}_2He^4 + {}_2He^4 + {}_2He^4.$$

7. Nuclear Reactions with Heavy Ions

Nuclear reactions with so-called heavy ions are those in which particles heavier than α-particles are used as projectiles. These are divided as :

(a) *Fission spallation and fragmentation reactions,* especially of the heavier elements are included in the third category. Mention may also be made of another kind of reaction which has been observed with target nuclei of moderate mass number, *e.g.,* the products of the interaction of 125 MeV nitrogen ions with aluminium 27 range from C^{12} to K^{38}.

$${}_{13}Al^{27} + {}_7N^{14} \rightarrow ({}_{20}Ca^{41})^* \rightarrow {}_1H^1 + 2{}_0n^1 + {}_{19}K^{38}.$$

Similar reactions, perhaps involving the emission of alpha particles, can lead to ${}_{14}Si^{31}$ and other nuclei in the same mass region.

(b) The first type, called *transfer reaction,* involves the transfer of a nucleon or a few nucleons from one nucleus to the other. The examples are :

$${}_7N^{14} + {}_5B^{10} \rightarrow {}_7N^{13} + {}_5B^{11} \rightarrow {}_7N^{13} + {}_7N^{15}$$
$${}_7N^{14} + {}_7N^{14} \rightarrow {}_6C^{13} + {}_8O^{15} \rightarrow {}_{13}Al^{25} + {}_7N^{16}$$
$${}_{13}Al^{27} + {}_7N^{14} \rightarrow {}_{12}Mg^{27} + {}_8O^{14}.$$

These reactions are known as *direct reactions* as they do not form any compound nucleus.

(c) In the second class of reactions with accelerated heavy ions, the compound nucleus is formed and two, three, four or more particles, mostly neutrons sometimes accompanied by one or two protons or α-particles are evaporated from the compound nucleus. The higher the energy of the projectile, the larger the number of particles expelled. The examples are :

Nb^{93} (C^{12}, 2n) Ag^{103}, Au^{197} (C^{12}, 4n) At^{205}, Al^{27} (C^{12}, αn) Cl^{34}, U^{238} (N^{14}, α5n) Bk^{243} and V^{51} (O^{16}, 2n) Ga^{65}.

INVERSE PROCESS

(Reciprocity Theorem): Let us consider a reversible process X + α = Y + y, in which X, x, Y and y occur in arbitrary numbers in a large box of volume V. We are interesting in the relation between the total cross-section σ(x → y), most generally σ(α → β) of the reaction with entrance channel α and reaction channel β and the total cross-section σ(β → α) of the inverse reaction. For this we use the fundamental theorem of statistical mechanics *(the principle of overall balance),* which states that when the system is in dynamical equilibrium all energetically permissible states are occupied with equal probability. Here we are interested in two particular states, the reaction channels α and β. The theorem is then equivalent to stating that in a given energy range the number of possible channels in the box is proportional to the number of possible channels into the box. The latter is given by

$$N_\alpha = \frac{4\pi p_\alpha^2 \, V \, dp_\alpha}{h^3} = \frac{p_\alpha^2 \, V \, dp_\alpha}{2\pi\hbar^3}.$$

Since $\quad \nu = \dfrac{dE}{dp}$, hence $N_\alpha = \dfrac{p_\alpha^2 \, V dE_\alpha}{2\pi^2\hbar^3 \nu_\alpha}$. ...(59)

Similarly, we have $N_\beta = \dfrac{p_\beta^2 \, V dE_\beta}{2\pi^2\hbar^3 \nu_\beta}$. ...(60)

The energy range for the two channels must of course be the same, *i.e.,* $dE_\alpha = dE_\beta$, hence

$$\frac{\text{No. of channels } \alpha \text{ in the box}}{\text{No. of channels } \beta \text{ in the box}} = \frac{N_\alpha}{N_\beta} = \frac{p_\alpha^2 \, \nu_\beta}{N_\beta \, \nu_\alpha}. \qquad ...(61)$$

The system is in dynamical equilibrium when the number of the transitions $\alpha \to \beta$ per second is equal to the number of transitions $\beta \to \alpha$ per second. The condition usually holds and is known as the *principle of detailed balance.* Further

No. of transitions $\alpha \to \beta$ per sec $= N_\alpha \times \omega(\alpha \to \beta)$,

where $\omega(\alpha \to \beta)$ is the transition probability for the reaction $(\alpha \to \beta)$.

Hence $$p\alpha^2\, v_\beta\, \omega(\alpha \to \beta) = p\beta^2\, v_\alpha\, \omega(\beta \to \alpha). \quad ...(62)$$

The transition probability measures the chance that one particle moving with velocity v in volume V is scattered per sec. Hence the cross-section σ which corresponds to unit incident flux is given by the relation

$$\sigma = \frac{\omega V}{v}. \quad ...(63)$$

Combining relations (62) and (63) and using $k = \frac{p}{\hbar}$, we have

$$k\alpha^2\, \sigma(\alpha \to \beta)\, k\beta^2\, \sigma(\beta \to \alpha) \quad ...(64)$$

or $$\frac{\sigma(\alpha \to \beta)}{\lambda\alpha^2} = \frac{\sigma(\beta \to \alpha)}{\lambda\beta^2}. \quad ...(65)$$

We have assumed zero intrinsic angular momenta for the particles so far. If I is the intrinsic angular momentum of any one of the particles, the corresponding density of states then must be multiplied by 2I + 1. This if there are intrinsic momenta for X, x, Y and y, we may write

$$(2I_X + 1)(2I_x + 1)\, k_\alpha^2 \sigma(\alpha \to \beta) = (2I_Y + 1)(2I_y + 1)\, k_\beta^2 \sigma(\beta \to \alpha). \quad ...(66)$$

If the initial and final states have definite angular momenta then the above equation must be employed.

CROSS-SECTION OF A NUCLEAR REACTION

Bohr assumed that the mode of disintegration of a compound system depends only on its energy, angular momentum and parity, but not on the specific way in which it has been produced. Therefore, to calculate the cross-section of a nuclear reaction, it is necessary to determine the cross-section of the two processes, the formation of the compound nucleus and its decay.

$$X + x \rightarrow C \rightarrow Y + y. \qquad ...(67)$$

Since the break up process of the compound nucleus is independent of the formation process, hence we can write in general

$$\sigma(\alpha, \beta) = \sigma_C(\alpha)\, G_C(\beta), \qquad ...(68)$$

where $\sigma(\alpha, \beta)$ is the cross-section corresponding to a specific entrance channel α and specific exit channel β, $\sigma_C(\alpha)$ is the cross-section for the formation of a compound nucleus C with particle x through channel α and $G_C(\beta)$ is the probability that the compound nucleus C will disintegrate by emitting a particle y through channel β.

For the sake of simplicity we shall neglect the dependence of $G_C(\beta)$ on angular momentum and parity. $G_C(\beta)$ then only depends on the excitation energy (E_C) of compound nucleus. Because of the Heisenberg uncertainty principle this energy cannot be sharply defined.

The uncertainty in excitation is measured by a quantity Γ_β, the level width associated with the decay constant λ_β of compound system.

$$\Gamma_\beta \text{ (For } E_C) = \hbar\lambda_\beta = \hbar\tau_\beta \text{ (for } E_C), \qquad ...(69)$$

where τ_β is the decay time of the compound nucleus through the channel β. If we add all the decay constants for the possible exist channels then we get the total decay constant for the compound nucleus. Therefore, the total width of the compound level with an excitation energy E_C is given by

$$\Gamma(\text{for } E_C) = \sum_\beta \Gamma_\beta(\text{for } E_C). \qquad ...(70)$$

The *branching probability,* the relative probability for the compound nucleus at excitation energy E_C to break up by emitting particle y, in terms of decay rates is given by

$$G_C(\beta) = T_\beta \text{ (for } E_C)/T(\text{For } E_C). \qquad ...(71)$$

The assumption of the independence of two processes (the creation and the disintegration of the compound system) gives a relation, which depends only in the kα and kβ. It is derived earlier and can be rewritten as

$$k\alpha^2\, \sigma(\alpha, \beta) = k\beta^2\, \sigma(\beta, \alpha). \qquad ...(72)$$

Using eqns (68) and (71), this relation becomes

$$k\alpha^2 \frac{\sigma_C(\alpha)\, \Gamma_\beta(\text{for } E_C)}{\Gamma(\text{for } E_C)} = k\beta^2 \frac{\sigma_C(\beta)\, \Gamma_\alpha(\text{for } E_C)}{\Gamma(\text{for } E_C)}$$

or
$$\frac{k\alpha^2\sigma_C(\alpha)}{\Gamma_\alpha(\text{for } E_C)} = \frac{k\beta^2\sigma_C(\beta)}{\Gamma_\beta(\text{for } E_C)} = U(E_C). \quad ...(73)$$

This proportional coefficient $U(E_C)$ is a function depending upon the excitation energy E_C of the compound nucleus only, but not on the particular channel used. This relation shows that there is a simple relationship between the probability for breakup and the cross-section for forming the compound nucleus through the same channel. The probability of decay through a channel β

$$G_C(\beta) = \frac{\Gamma_\beta}{\Gamma} = \frac{k\beta^2\sigma_C(\beta)}{[\Gamma_\alpha + \Gamma_\beta + ...]U(E_C)} = \frac{k\beta^2\sigma_C(\beta)}{\sum_\gamma k_\gamma^2\sigma_C(\gamma)}, \quad ...(74)$$

where the sum is extended over all channels γ into which compound system can decay. It is, therefore, possible to derive the decay probability through a given channel of the compound nucleus if the cross-section $\sigma_C(\gamma)$ for the formation of the compound nucleus by all possible channels are known, and if the Bohr assumption is valid. The validity of Bohr's hypothesis of the compound nucleus has been elegantly demonstrated by several experiments. Let us discuss the experiment of Indian Scientist S.N. *Ghoshal* (1950) in which he produced the same compound nucleus Zn^{64*} with the bombardment of $_{28}Ni^{60}$ by α-particle and $_{29}Cu^{52}$ by protons. Additional 7 MeV was added to the K.E. of protons to produce the same excitation in Zn^*. The reactions observed were

Ni^{60} (α, n) Zn^{63} $\qquad$ Cu^{63} (p, n) Zn^{63}

Ni^{60} (α, 2n) Zn^{62} $\qquad$ Cu^{63} (p, 2n) Zn^{62}

Ni^{60} (α, pn) Cu^{62} $\qquad$ Cu^{63} (p, pn) Cu^{62}.

Since the excitation produced through these two processes are the same, the disintegration probability $G_C(\beta)$ is the same, because it depends upon the excitation produced in the compound nucleus and not upon the mode of formation. If the compound nucleus theory is true, then one should expect.

$$\sigma(p, n) : \sigma(p, 2n) : \sigma(p, pn) = \sigma(\alpha, n) : \sigma(\alpha, 2n) : \sigma(\alpha, pn). \quad ...(75)$$

Measurements due to Ghoshal, here six cross-sections are plotted as functions of K.E. of the β-particle and protons. The proton-energy scale has been shifted by 7 MeV with respect to the α-particle energy scale in order to make the peaks of the proton curves correspond with those of the α-particle curve. It can be seen from figure that the results are accurate with in 10%, and thus providing evidence for the validity of the compound nucleus theory. The small difference between experimental and theoretical values can be explained as under : Ghoshal's experiment was performed at high bombarding energies. The levels excited in the compound nucleus are more closely spaced and broader and may overlap partially. This can be explained by the *continuum theory.* Thus we see that the compound nucleus theory given here is applicable to low bombarding energy.

CONTINUUM THEORY OF NUCLEAR REACTION

At higher bombarding energies the individual levels of the compound nucleus become broader and also more closely spaced. The total width Γ becomes much greater than D, the spacing of the levels. The width and number of levels are such that even though the interval is small, many levels may be found in it. Sharp resonances are no longer observable when $\Gamma \geq D$. The spacing between levels is completely occupied, so the *space is described as continuum.* The cross-section for the formation of a compound nucleus is larger for neutrons than for charged particles because Coulomb repulsion between the incident charged particle and the target nucleus is important in the latter case. In the medium and heavy nuclei, the individual level becomes broader and levels become more closer when the energy of the incident particle is large. The continuum theory of nuclear reaction cross-sections treats the individual level not separately but as an average over many resonances.

We shall derive the value of cross-section on the basis of classical mechanics. If the incident particle is a neutron (with J = 0), the cross-section for reaching the nuclear surface is given by the classical target area such as,

$$\sigma_C(n) = \pi R^2. \qquad \text{...(76)}$$

We now proceed to a qualitative wave mechanical discussion. This discussion is based upon the fact that there is a sudden change of potential when the particle crosses the boundary of the nucleus, due to the strongly attractive nuclear forces. The sudden change of potential reflects the incoming wave at the nuclear surface and thus decreases the cross-section, especially at low energy.

Let us consider a beam of particles of energy E moving in the direction of x in a potential V(x), which is zero for x < 0 and is $-V_0$ for x > 0. The particles are partially reflected at x = 0. We have to determine now the transmission coefficient T, defined as the ratio of the number of particles transmitted through the region x > 0 to the number of incident particles. Thus

$$\text{For } x < 0,\ \psi = Ae^{ikx} + Be^{ikx}, \qquad ...(77)$$

where $$k = \sqrt{\frac{(2ME)}{\hbar}}. \qquad ...(78)$$

$$\text{For } x > 0,\ \psi = Ce^{+iKx} \qquad ...(79)$$

where $$K = \sqrt{\frac{[2M(E+V_0)]}{\hbar}} \qquad ...(80)$$

Here A, B and C are coefficients giving the amplitude of the various wave functions. Since neither the wave function nor the kinetic energy can become infinite at any point, hence both, the wave function and its derivative must be continuous at all points. Using boundary conditions at x = 0, we have

A + B = C and ikA – ikB = iKC.

$$\therefore\ A = \frac{1}{2}\left(\frac{1+K}{k}\right) \text{ and } B = \frac{1}{2}C\left(\frac{1-K}{k}\right) \qquad ...(81)$$

Hence $$T = \frac{|A|^2 |B|^2}{|A|^2} = \frac{4Kk}{(K+k)^2} \qquad ...(82)$$

Hence limiting values of T are of special interest. At high incident energies relation (80) indicates that K = k and thus T approaches 1. On the other hand, as the neutron energy approaches zero,

$$k < K \text{ and } T = \simeq \frac{4k}{K} \simeq 4\left(\frac{E}{V_0}\right)^{1/2}.$$

The position of the particle is uncertain by the amount of its wavelength $\lambda\!\!\!^{-}$. This can be accounted approximately for by replacing πR^2 with $\pi(R + \lambda\!\!\!^{-})^2$. Thus cross-section becomes

$$\sigma_c = \pi(R + \lambda\!\!\!^{-})^2. \qquad ...(83)$$

This equation has to be modified by including the transparency T of the potential barrier. Thus we obtain the cross-section for the formulation of the compound nucleus by S-wave neutrons.

$$\sigma_c = \pi(R + \lambda\!\!\!^{-})^2 4kK/(k + K)^2. \qquad ...(84)$$

For very small neutron energies k ≤ K and $\lambda\!\!\!^{-} \geq R$, the above equation becomes

$$\sigma_c = \frac{4\pi}{kK} \cdot \left(\text{Because } k = \frac{1}{\lambda\!\!\!^{-}}\right) \qquad ...(85)$$

At high energies the cross-section approaches the classical value πR^2 and increases with decreasing energy, but less strongly than $\pi(R + \lambda\!\!\!^{-})^2$, because of the reflection at the nuclear boundary.

Similar approach can also be used to obtain an expression for the reaction cross-section of charged particles. In this case the incident beam is deviated by the Coulomb potential $V(r) = Z_1Z_2e^2/2\pi\epsilon_0 r$. The incident particles reaching the nuclear surface thus have a maximum impact parameter $R[1 - (B/E)]^{1/2}$, where B = Coulomb barrier height = V(R). Hence the reaction cross-section can be expressed as

$$\sigma_c = \pi R^2[1 - B/E] \qquad E > B$$

$$= 0 \quad E \leq B. \qquad ...(86)$$

Here we see that the high energy limit of this eqn. is πR^2, identical to that for neutrons.

An attempt to obtain a better fit to the continuum theory value of σ_c has been made by Destrovsky *et al.*, in 1959.

THERMONUCLEAR REACTIONS

Power from nuclear fission is now a reality both on land and sea and in those countries where coal or oil is costly. An alternative to the fission reaction as a source of energy is its reverse process known us *fusion process* in which the lighter nuclei fuse together and produce a heavier nucleus. The sum of the masses of the Individual light nuclei is more than would be the mass of the nucleus formed by their fusion, and thus the fusion process should result in a liberation of energy.

Some indication of how this might be achieved can be obtained by considering the source of energy produced continuously in the stars, including the sun. It has been calculated that the sun (our nearest star) emits electromagnetic energy at a rate of about (our joules per second. Astronomical and geological evidences show that the sun has been radiating energy at about its present rate for several billion years. Chemical reactions cannot possibly be the source of this energy because even if the sun is supposed to be consisted of pure carbon, its complete combustion would supply energy to maintain these radiations only for few thousand years.

Now question arises how can the sun have maintained this energy output for so long and what is the source of all *stellar energy*? Helmholtz in 1853, suggested that the contraction was taking place and thus the gravitational energy was being converted into heat energy in sun. This conversion is analogous to the production of electricity from the falling of water. It has been shown that if contraction were taking place it could supply not more than 1% of the total energy output needed and it led to an estimate of the age of the sun which was much too short.

With the discovery of radioactivity at the end of the 19th century, it appeared possible that atomic energy might be contributing to the sun's energy. In 1904, J.II. Jeans suggested that the energy of the sun might be resulted from the mutual annihilation of +ve and –ve charges. Eddington in 1920, suggested that the stellar energy was liberated in the formation of helium from hydrogen. This theory received wide support, although there was no satisfactory mechanism to account for the formation of helium from hydrogen because large amounts of hydrogen and helium exist in the sun. In 1929, Atkinson and Houtermans, in Germany, considered that energy might be liberated in the very high stellar temperatures. Such processes are called *thermonuclear reactions*.

In order to interact two nuclei, that must have enough kinetic energy to permit them to overcome the electrostatic repulsion barrier which tends to keep them apart. It can be shown from calculations that the energy requited make the nuclear reactions occur at a detectable rate is about 0.1 MeV for nuclei of the lowest atomic number (*e.g.,* isotopes of hydrogen), the larger energies are needed for nuclei of higher atomic number. This energy can be resulted from a sufficient increase in temperature (1000×10^6 °K if the average energy of the particles is to be 0.1 MeV). Such temperatures are considerably higher than those existing in stars from stars, like sun where the central temperatures are less than (50×10^6 °K, the fusion reaction takes place with a finite rate and releases enough energy to keep up the heat and light of the star.

H.A. Bethe in the United States suggested in 1939 that the *production of stellar energy is by thermonuclear reactions* in which helium-4 nuclei are synthesized from four protons (the nucleus of hydrogen, the most abundant element in the universe). A few years ago, it was held that the major portion of the sun energy was derived from the *carbon nitrogen cycle*. Recent modification of the estimates of the central temperature of the sun now favour the *proton-proton chain*. In the carbon-nitrogen cycle carbon acts as a short of catalyst in facilitating the combination of four protons to form a helium nucleus. In this cycle a proton first interacts with a C^{12} nucleus with a release effusion energy. The product nucleus N^{13} decays in a very short time. The stable nucleus of C^{13}, thus formed then reacts with another proton. The more energy being liberated by this process. The stable nucleus of N^{14} combines with a third proton. The product nucleus O^{15} is a positive β emitter, which decays into N^{15}. This nucleus finally interacts with a fourth proton and regenerates C^{12} nucleus. These reactions can be written as:

$${}_6C^{12} + {}_1H^1 \rightarrow ({}_7N^{13}) \rightarrow {}_7N^{13} + 1.94 \text{ MeV } (10^6 \text{ y})$$

$${}_7N^{13} \rightarrow ({}_6C^{13}) + {}_{+1}e^0 + \nu + 1.20 + 1.02 \text{ MeV } (10 \text{ m})$$

$${}_6C^{13} + {}_1H^1 \rightarrow ({}_7N^{14}) \rightarrow {}_7N^{14} + 7.55 \text{ MeV } (2 \times 10^5 \text{ y})$$

$${}_7N^{14} + {}_1H^1 \rightarrow ({}_8O^{15}) \rightarrow {}_8O^{15} + 7.29 \text{ MeV } (3 \times 10^7 \text{ y})$$

$${}_8O^{15} \rightarrow {}_7N^{15} \rightarrow {}_{+1}e^0 + \nu + 1.74 + 1.02 \text{ MeV } (2\text{m})$$

$${}_7N^{15} + {}_1H^1 \rightarrow ({}_8O^{16}) \rightarrow {}_6O^{12} + {}_2He^4 + 4.96 \text{ MeV } (10^4 \text{ y})$$

$$4\ {}_1H^1 \rightarrow {}_2He^4 + 2\ {}_{+1}e^0 + 2\nu + 26.7 \text{ MeV}$$

It will be noted this chain of reactions can start with either carbon or nitrogen, since each one is reproduced in the reaction. The four protons are associated with four elections to maintain electrical neutrality, whereas only two are required for the helium nucleus, and rest two electrons combine readily with positrons resulting in the formation of γ-rays.

The mass difference released as energy in this chain of reactions is simply the difference between the masses of four protons and one helium nucleus. A small amount of this energy is carried away by the neutrons that are emitted during the e^+-decay. Bethe estimates this to be about 1.84 MeV leaving rest for each α-particle formed.

In the p-p chain, two protons first fuse to produce a deuterium nucleus which combines with an another proton to yield He^3. Two He^3 nuclei interact and form He^4 and two protons. These reactions can be represented by the equations

$$
\begin{aligned}
{}_1H^1 + {}_1H^1 &\rightarrow ({}_2He^2) \rightarrow {}_1H^2 + {}_{+1}e^0 + \nu + 0.42\text{ MeV } (7 \times 10^9\text{ y})\\
{}_1H^2 + {}_1H^1 &\rightarrow ({}_2He^3) \rightarrow {}_2He^3 + \gamma + 5.5\text{ MeV } (10\text{ sec})\\
{}_2He^3 + {}_2He^3 &\rightarrow ({}_4Be^6) \rightarrow {}_2He^4 + {}_1H^1 + {}_1H^1 + 12.8\text{ MeV } (3\times10^5\text{ y})\\
\hline
4\ {}_1H^1 &\rightarrow {}_2He^4 + 2\ {}_{+1}e^0 + 2\nu + 2\gamma + 26.7\text{ MeV}
\end{aligned}
$$

The positrons emitted are annihilated by free electrons with the production of gamma rays. The energy released in p-p chain is the same as in the C-N cycle (26.7 MeV for each helium nucleus)

Which of the two hydrogen-helium fusion processes plays the major role in energy production in star? The answer is based on our knowledge of the stellar temperature. H. Bondi and E.E. Salpeter (1952) developed empirical equations for the rate at which energy is liberated in each of the above chains of reaction. The p-p chain reaction rate varies more slowly with temperature, roughly as T^4 and is much more important at lower temperatures. The carbon cycle rate predominates as the temperature is in the vicinity of 18×10^6 °K.

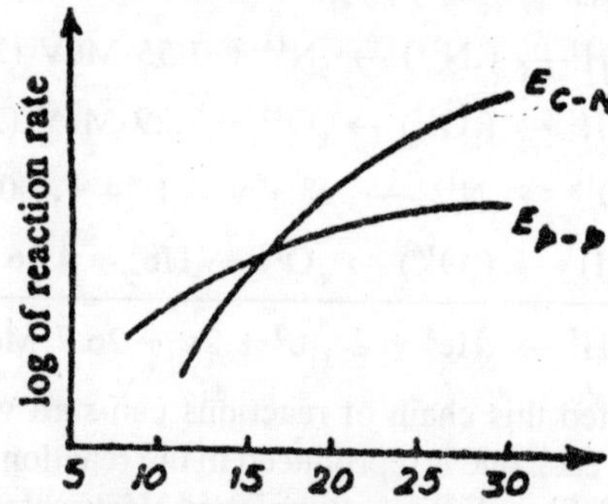

Fig. 1 : Variation in the rate of energy produced with the central teftiperature.

For the case of the sun, they estimated that the rate of generation of energy in the p p chain was about the same as that in the carbon cycle. Actually the p-p chain predominates (96%) because the interior temperature of the sun is about 15×10^6 °K.

Solar Future

When all the hydrogen is used up in the above thermonuclear reactions, the star will consist mostly of helium. At this stage gravitational contraction will occur once again until a temperature of about 10^8 °K is reached and the density of the star is about 10^7 kg/m^3. Under these conditions three helium nuclei combine and form a C^{12} nucleus with the release of about 7.3 MeV. Hoyle estimates that such processes provide energy for an additional 10^7 years. Further gravitational contraction of the star will occur, when all the helium is used up, and will produce a further rise in the temperature of the star. The atoms formed in this way are most stable. Any appreciable combination of these atoms to form heavier ones will lead to endothermic rather than exothermic reactions. Bondi and Salpeter suggested that the endothermic reactions might account for the sudden collapse of a star identified as the sudden appearance of a supernova.

RESONANCE : BREIT WIGNER DISPERSION FORMULA

The concept of cross-section and level width can be applied to resonances in a quantitative way. In the particularly important case of resonance processes, a theoretical formula for the cross-section was derived by G. Breit and E.P., Wigner in the United States in 1936. In its simplest form, it gives the value of the cross-section in the neighbourhood of a single resonance level formed by an incident particle with zero angular momentum and charge zero so that spin and the Coulomb effects can be ignored. The result is analogous to the theory of optical dispersion, so that the main formula obtained is often called the *dispersion formula.*

Whether or not the level of compound nucleus is bound, excitation by an incident particle can be treated as analogous to the excitation of the oscillations produced in an electric circuit by an electromagnetic wave. We therefore, expect the nuclear cross-section to vary with the incident energy in the same way that the energy in a forced oscillation varies with incident frequency. The classical resonant circuit absorbs energy because of resistive levels. In the nuclear case, damping arises because of the possibility of decay. Because of this possibility the nuclear state has a finite width Γ. The wave function of a decaying state of mean energy E_0 may be written as

$$\psi(r,t) = \psi(r)e^{iE_0 t/\hbar}e^{-\Gamma t/2\hbar}. \qquad ...(87)$$

This corresponds to an exponential decrease of intensity of excitation $|\psi(r, t)|^2$ with a time constant $\tau(=\hbar/\Gamma)$. This wave function also shows that a decaying state is not a state of definite energy E of the form

$\Psi(r)e^{-i(E/\hbar)t}$. Never the less, it can be represented by superposition of states of slightly different energies E, each with a different amplitude A(E)

$$\Psi(r, t) = \int_{-\infty}^{+\infty} A(E)e^{-iEt/\hbar}dE. \qquad ...(88)$$

Using the Fourier analysis technique, we can show that the energies E are grouped about a mean energy E_0 with a spread of the order of $\Gamma = \hbar\lambda$. Equating eqns. (87) and (88), we get

$$e^{-\Gamma t/2\hbar} = e^{-\lambda t/2} = \int_{-\infty}^{+\infty} A(E)e^{-iEt/\hbar}dE. \qquad ...(89)$$

According to the Fourier theorem any well behaved function f(t) can be represented as

$$f(t) = \frac{1}{2\pi}\operatorname*{Lim}_{s\to\delta}\int_{-\Omega}^{\Omega} - i\omega t \, d\omega \int_{-\infty}^{\infty} e^{j\omega t'} f(t')dt' \qquad ...(90)$$

Applying this to the function $e^{-\lambda_t/2}$, we get

$$4(E) = \frac{1}{2\pi\hbar}\int_0^{\infty} e^{\left[i(E-E_0)/\hbar - \lambda/2\right]t'} dt'$$

$$= \frac{1}{2\pi} \frac{1}{(E-E_0) + i\hbar\lambda/2}. \qquad ...(91)$$

Here we have assumed that the decaying system was prepared at the time t = 0. The probability of finding the system with a given energy E is proportional to

$$|A(E)|^2 = \frac{1}{4\pi^2} \frac{1}{(E-E_0)^2 + (\hbar\lambda/2)^2}$$

$$\frac{1}{4} = \frac{1}{4\pi^2} \frac{1}{(E-E_0)^2 + \frac{\Gamma^2}{4}}. \qquad ...(92)$$

This gives the level shape. It is exactly as for pure radiative decay except that particle emission is now included by using the total width Γ instead of the radiative width Γ_r. The cross-section for excitation of the level by collision of particle x width nucleus X is, therefore, expected to have the form

$$\sigma_x = \frac{C}{\left[(E - E_0)^2 + \frac{\Gamma^2}{4}\right]}, \qquad ...(93)$$

where C is constant. For the determination of C, let us suppose that the compound nucleus formation and decay processes take place in a box of volume V which contains one nucleus X and one particle x. If the states are quantised, the no. of states of motion of particles with momentum between p and p + dp = $(4\pi p^2 dp V)/h^3$.

The probability of formation of the compound level per unit time = No. of possible states of motion × probability that the nucleus X is contained within the small volume σ_{xv} out by the

$$dP = \frac{4\pi p^2 dp}{h^3} V \cdot \frac{\sigma_{xv}}{V} = \frac{4\pi}{h^3} = \nu \sigma_x p^2 dp.$$

Substituting the value of σ_x from eqn. (93) in the above eqn. and integrating over the energy spectrum we get

$$P = \int_{-\infty}^{+\infty} \frac{4\pi}{h^3} \nu \cdot p^2 \cdot \frac{C}{(E - E_0)^2 + \frac{\Gamma^2}{4}} dt$$

$$= \int_{-\infty}^{+\infty} \frac{4\pi}{h^3} \cdot \frac{1}{(E - E_0)^2 + \frac{\Gamma^2}{4}} dE \quad .$$

If we assume that the variation of the channel wavelength λ of the particle over the level width Γ may be neglected, then the probability of formation per sec

$$= \frac{4\pi C}{h\lambda^2} \cdot \frac{2}{\Gamma} \left[\tan^{-1} \frac{2(E - E_0)}{\Gamma} \right]_{-\infty}^{+\infty}$$

$$= \frac{4\pi C}{h\lambda^2} \cdot \frac{2}{\Gamma} \cdot \pi = \frac{C}{\hbar \pi \lambda\!\!\!^{-2} \Gamma}. \qquad ...(94)$$

Probability of decay per sec $= \dfrac{\Gamma_x}{\hbar}$,

where Γ_x is the partial width of the compound level for the emission of x. In the equilibrium state the probability of formation per sec would equal to the probability of decay per sec of the excited state back into the system X + x. Hence

$$\frac{C}{\pi\hbar\lambda\!\!\!^{-2}\Gamma} = \frac{\Gamma_x}{\hbar} \text{ or } C = \pi\lambda\!\!\!^{-}\Gamma\Gamma_x, \qquad ...(95)$$

Substituting this value in eqn. (93), the cross-section for the formation of the level becomes

$$\sigma_x(E) = \frac{\pi\lambda\!\!\!^{-2}\Gamma\Gamma_x}{(E-E_0)^2 + \frac{\Gamma^2}{4}} \qquad ...(96)$$

For the process X(x, y) Y, we obtain reaction cross-section as

$$\sigma_r = \sigma_{xy}(E) = \frac{\sigma_x\Gamma_y}{\Gamma} = \pi\lambda\!\!\!^{-2}\frac{\Gamma_x\Gamma_y}{(E-E_0)^2 + \frac{\Gamma^2}{4}}. \qquad ...(97)$$

Here Γ_x and Γ_y are partial level width defined as

$$\Gamma_{x\tau x} = \hbar \text{ and } \Gamma_{v\tau v} = \hbar$$

where τ_x and τ_v the main life times that the compound nucleus would have if elastic scattering of x or the emission of y were the only possible modes of decay.

If spin is considered, the right band side of eqn. (97) must be multiplied by the factor

$$gc = \frac{(2Ic+1)}{(2I_x+1)(2Ix+1)}, \qquad ...(98)$$

where I_x is the total angular momentum of the incident particle, Ix is that of target nucleus and Ic is of the compound state, which is formed only by those orbital angular momentum l_x which satisfy the conditions

$$Ic + Ix + I_x + I_x \text{ and } \Pi x \Pi_x (-1)^{lx} = \Pi c. \qquad ...(99)$$

Hence for the nuclear reaction in which particles have definite spins, the relation (97) can be written as

$$\sigma_r = \pi\lambda\!\!\!^{-2}\frac{(2Ic+1)}{(2I_x+1)(2Ix+1)}\frac{\Gamma_x\Gamma_y}{(E-E_0)^2 + \frac{\Gamma_v^2}{4}} \qquad ...(100)$$

This is known as the *Breit Winger resonance formula.* For the (n, γ) reaction in particular

$$\sigma(n,\gamma) = \pi\lambda\!\!\!^{-2} = \frac{(2Ic+1)}{2(2Ix+1)}\frac{\Gamma_x\Gamma_y}{(E-E_0)^2 + \frac{\Gamma^2}{4}} \qquad ...(101)$$

This is maximum when $E = E_0$ and is equal to

$$\sigma_{max}(n, \gamma) = 4\pi ƛ^2 gc\Gamma_n\Gamma_\gamma/\Gamma^2. \qquad ...(102)$$

For $E = E_0 \pm \Gamma/2$, $\sigma = 1/2\ \sigma_{max}$ and hence Γ is the full width at half maximum. A sharp resonance corresponds to a narrow width Γ and thus to an excited state of long life. A largest possible capture cross-section will occur when $\Gamma_n = \Gamma_\gamma = \Gamma/2$. Its maximum possible value $\sigma_{max} = \pi gc ƛ'^2$.

The width of the resonance peak affects the cross-section. In a general way, if the peak is broad, covering a large energy range, the cross-sections are likely to be somewhat decreased as compared with the case of sharp and narrow peak. The width of the peak is inversely related with the life of the excited state of the compound nucleus.

The total cross-section σ_t and the radioactive capture cross-section $\sigma(n, \gamma)$ have been measured as functions of the incident neutron energy. Some of the results are which these cross-sections are plotted as functions of energy in the intermediate range (10 – 500 keV). In this energy range total cross-section for Al^{27} is practically equal to the elastic scattering cross-section as only reactions in this case are the elastic scattering and radioactive capture.

Equation (100) applies to all resonance cross-sections except for the elastic re-emission of the incident particle σ(x, x), *i.e., elastic resonance scattering.* Besides compound nucleus scattering (*i.e.,* absorption and re-emission of neutron of the same energy), the incident particle wave is scattered by the nucleus as if it were an impenetrable sphere. This type of scattering is known as *potential* or *shape-elastic scattering.* The elastic-scattering cross-section σ_{el} is given by

$$\sigma_{el} = 4\pi ƛ^2\left[gc\left|\frac{\Gamma_{n/2}}{\left(E - E_0 + \frac{i\Gamma}{2}\right)} + e^{i\phi_l}\sin\phi_l\right|^2 + (1 - gc)\sin^2 f_l\right], \qquad ...(103)$$

where ϕ_l is an energy dependence quantity known as the hard sphere phase shift. For $l = 0$ or S-wave neutrons $\phi = R/ƛ = Rk$. Hence for S-wave neutrons on a spinless target

$$\sigma_{el} = 4\pi ƛ^2\left|\frac{\Gamma_{n/2}}{(E - E_0) + i(\Gamma/2)} + e^{kR}\sin kR\right|^2. \qquad ...(104)$$

Here the coefficient $4\pi\lambda\!\!\!^{-2}$ is the maximum possible S-wave scattering cross-section. The first term between bars is called the resonance scattering amplitude, which if present alone would lead to eqn. (97). The second term is called the *potential scattering amplitude,* which if present alone would lead to

$$\sigma_{el} = 4\pi\lambda\!\!\!^{-2} \sin^2 kR. \qquad ...(106)$$

This term varies smoothly with energy. For small energies $kR \leq 1$. As $E \to 0$, we get

$$\sigma_{el} \simeq 4\pi R^2, \qquad ...(106)$$

which is the potential scattering cross-section for an impenetrable sphere when $R \leq \lambda\!\!\!^{-}$.

The resonance term rises to large values near $E = E_0$, but is small elsewhere. For $E < E_0$ the two terms interfere destructively yielding a low value of σ_{el}. For $E > E_0$ the interference is constructive. For charged particles, resonance scattering involves coherence between Coulomb potential scattering, nuclear potential scattering and resonance scattering.

Let us consider the lowest energy resonance of the compound nucleus. For $E \leq E_0$, the denominator of eqn. (101) does not change much with E and Γ_γ is independent of E but Γ_n does depend on E. Hence $\sigma(n, \gamma) \propto \lambda\!\!\!^{-2}\, \Gamma_n$. The probability of elastic emission of a neutron of energy E (momentum p) is proportional to the density of states in momentum space around p. Hence

$$\Gamma_n \propto \frac{p^2 dp}{dE} \alpha p \left\{ \begin{aligned} E &= \frac{p^2}{2m} \\ dE &= \left(\frac{p}{m}\right) dp \end{aligned} \right\}$$

Thus we have $\sigma(n, \gamma) \propto \lambda\!\!\!^{-2} p$

$$\propto \frac{1}{p} \qquad \left(\because \lambda\!\!\!^{-} = \frac{\hbar}{p}\right)$$

$$\text{or} \qquad \propto \frac{1}{v}, \qquad ...(107)$$

where v is the velocity of neutron. *It is known as l/v – law.* Thus Breit-Wigner formula leads to the conclusion that at low neutron energies the cross-section should be inversely proportional to the neutron speed.

Following 1/v *region* there occurs the *resonance region* in which the cross-sections rise sharply to high values. The cross-sections are low with neutrons of very high energy.

OPTICAL MODEL

The continuum theory does not stand upto experimental tests satisfactorily. Barschall plotted the measured total neutron cross-sections in a three dimensional graph against the neutron energy E and the mass number A, and found that the cross-sections did not decrease smoothly with increasing E, as predicted by the theory, and that the trend of those maxima and minima with energy was a smooth function of A. He showed that the disagreement with the theory was not due to unexpected resonances in individual nuclides, but to a general flow in our theory. The widely spaced shallow maxima and minima were expected from scattering by a potential well, as in the shell model. Thus we see that *our compound nucleus theory must be modified in the light of shell model idea.*

A mathematical mode, known as *optical model,* pictures the interaction among the nucleus as being intermediate to that predicted by the continuum and shell models of the nucleus. It is the model of a complex nuclear potential, the real part produces a potential scattering like the scattering by a hard sphere and imaginary part corresponds to the cross-section for compound nucleus formation.

The nucleus can thus be viewed as a *cloudy crystal ball,* which is able to transmit, refract and absorb the incident particle waves, analogous to the transmission, reflection, refraction and absorption of a light beam. In optics we require an index of refraction which is a complex quantity to describe optical phenomena. Due to this analogy with optics, this nuclear model is named as *optical model.*

Let us first discuss the optical model of Fernbach, Serber and Taylor, which involves a nuclear potential of complex form. For the sake of simplicity, we consider the case that the real and imaginary parts of the nuclear potential have the same radial dependence, such as

$$\begin{aligned} V &= V(r)\ (1 + i\xi) && \text{for } r \le R \\ V &= 0 && \text{for } r \ge R. \end{aligned} \qquad \text{...(108)}$$

The Schrodinger equation of the neutron with this potential is

$$\nabla^2\psi + \frac{2M}{\hbar^2}\ [E - V(r)\ (1 + i\xi)]\ \psi = 0. \qquad \text{...(109)}$$

For the special case of $l = 0$ neutrons and a square well of depth V_0, for which

$$V(r) = -V_0\ (r < R),\ V(r) = 0\ (r > R),$$

the radial equation is given by

$$\frac{du^2}{dr^2} + K_c^2 u = 0, \quad r < R \qquad \text{...(110)}$$

and $$\frac{d^2u}{dr^2} + k^2 u = 0 \quad r > R, \qquad \text{...(111)}$$

where

$$Kc = \frac{\sqrt{[2M(E + V_0 + i\xi V_0)]}}{\hbar} = \left(\frac{2M}{\hbar^2}\right)^{1/2} (E + V_0)^{1/2} \left(1 + \frac{i\xi V_0}{E + V_0}\right)^{1/2}$$

$$= \frac{[2M(E + V_0)]}{\sqrt{\hbar + i\xi V_0 \left[\frac{M}{2\hbar^2 (E + V_0)}\right]^{1/2}}} = K + ik'$$

Hence the solution of eqn. (110) is of the form

$$\psi(r) = \left(\frac{A}{r}\right) \sin Kcr$$

or $$|r\psi|^2 = A^2[\sin^2 Kr + \sinh^2 k'r), \text{ for } r \le R. \qquad \text{...(112)}$$

The wave is absorbed as it penetrates into the potential step. The characteristic length known as penetration depth for $E \le V_0$ is given by

$$L = \frac{1}{k'} = \frac{2\hbar(E + V_0)^{1/2}}{\xi V_0 \sqrt{(2M)}} = \frac{2\hbar}{\xi(2MV_0)^{1/2}} = \frac{2}{\xi K} \qquad \text{...(113)}$$

Substitution of experimental values ($x \approx 0.06$, $V_0 \approx 50$ MeV), we get L of the order of nuclear diameter. This model has explained discontinuities in the variation of the elastic cross-sections for slow neutrons as a function of the mass number of nuclei. The maxima has been observed for radii $R = \left(n + \frac{1}{2}\right) \lambda_{inside}$. The study of elastic scattering has provided valuable information on nuclei radii. The radii so calculated were not corresponding to the relation $R = R_0A^{1/3}$. The above potential form gives too large a cross-section for the elastic scattering process in backward directions.

A much better fit to the experimental cross-sections is the potential well proposed by Woods and Saxon. In it V(r) varies smoothly with r, the distance from the centre of the nucleus, and is of the form

$$V(r) = -(V_0 + iW_0)\,[1 + e^{(r - R)/a}] \qquad \text{...(114)}$$

The plot of its real part is shown in Fig. 2.14. The curve shows that the potential well is no longer square but has rounded edges. R is the radius of the equivalent square well and is very nearly equal to the radius at which the depth of the potential well is one-half its central value a is a surface density parameter and is a measure of the degree of the tapering of the potential near the nuclear surface. The optical model parameter V_0 and W_0 $(= \xi V_0)$ are dependent on the type and energy of the incident particle, V_0 decreases smoothly with the incident energy.

This model gives little idea about the nuclear surface. A better fit to the experimental results is obtained if a surface absorption potential of Gaussian shape, centred around r = R is included.

In recent years other forms of optical potentials have been introduced to take into account the deformation of the nuclei, the spin-orbit coupling and also the energy region at which the experiments are being performed. Additional terms are required to take into account these effects. *The real part the potential function accounts for refraction and imaginary part for absorption of the incident nucleons. The spin-orbit term describes changes in spin orientations through the coupling of spin and angular momenta.*

This orientation is known as polarization. In addition to the theory of the neutron-nuclear interaction, the optical model has been successively applied to the scattering of protons, deuterons, alpha particles and heavier ions. We *shall give few experimental results.* R.D. Albert in 1959 has measured (p, n) cross-sections in some medium weight nuclei between the energy range 4 to 5.5 MeV, and used the Bjorklund and Fernback optical potential which has the form

$$U(r) = V_\rho(r) + iWq(r) + \frac{\lambda\hbar^2}{4m_0^2c^2}\,\frac{V}{r}\,\frac{d\rho}{dr}\,(\vec{\sigma}.\vec{l}) \qquad \text{...(115)}$$

where $\rho(r) = [1 + e^{(r - R)/a}]^{-1}$, $q(r) = e^{-[(r - R)/b]^2}$, $R = r_0A^{1/3}$ and m_0 = mass of the electron.

The results of the measurements made by Albert. The theoretical curves, obtained with parameters of the following table, adopted as giving the best fits to experimental data are also indicated by the thick lines. These results show that there is an excellent agreement between theory and experiment.

E(MeV)	V(MeV)	W(MeV)
4.0	47	7.0
4.5	45	7.7
5.0	45	8.25
5.5	44	8.7

The optical potential function represented by equation (115) has also been used by Shore, Wall and Irvine to analyze angular distribution of 7.5 MeV protons scattered elastically by vanadium. They used two sets of parameters corresponding to surface absorption and to volume absorption. Recently, Rosen, Beery, Goldhaber and Auerbach have proposed a spin dependent optical potential by analyzing and measuring 80 separate angular distribution data.

Halbert, Bassel Satchler carried out an analysis of the d-d scattering data at 11 MeV and at 11.8 MeV. They used nuclear potentials of the form

$$V(r) = V\left[1+e^{\left(r-r_0A^{1/3}\right)/a}\right]^{-1} + iW\left[1+e^{\left(r-r_wA^{1/3}\right)/a_w}\right]^{-1} \quad ...(116)$$

for volume absorption and

$$V(r) = V\left[1+e^{\left(r-r_0A^{1/3}\right)/a}\right]^{-1} - ia_w W_0 \frac{d}{dr}\left[1+e^{\left(r-r_wA^{1/3}\right)/a_w}\right]^{-1}$$

for surface absorption.

In each case uniform charge distribution of radius 1.3 $A^{1/3}$ F was assumed for the Coulomb potential. They obtained an excellent fit for the target nuclei Ni, Zr, Ag and Sn and a reasonably good fit for Ca and Ti.

The case of the deuteron is complicated because, due to polarization or stripping of the deuteron by Coulomb interaction, non-elastic channels

The optical potential may contain spin-orbit terms when the incident particle has a spin other than zero. The strength of the spin-orbit potential is adjusted to fit the measured polarization of the scattered particles. It has been found from experiments that the idea of the optical potential is not limited to the calculation of cross-sections of neutron reactions and incident energies of a few million electron volts but is applicable to the description of reactions of protons and heavier particles in the region of high energies as well.

The basic idea of the shell model is present in the optical model analysis of nuclear reactions. If the imaginary parts of the potential were absent, we should have an ordinary Schrodinger equation and corresponding stationary orbits. The shell model shows that such an approximation has considerable merit.

DIFFERENT STAGES OF A NUCLEAR REACTION

Detailed theories of nuclear reaction were patterned after the two nuclear models. The rigid drop model and the shell model. In the *compound nucleus* theory it was assumed that a nuclear projectile incident on a nucleus would interact strongly with all the nucleons in the nucleus and quickly share its energy with them. In the *reaction theory* (optical model theory), it was proposed that an incident nucleon would interact with the nucleus via the shell model potential. The success of the optical model suggests that the Bohr theory of compound nucleus is in need of modification. A more general scheme of nuclear reaction this been described by Weisskopf. According to him the nuclear reaction proceeds through three stages : the *independent particle (optical model) stage, compound nucleus stage and the final stage.*

During the first stage, the interaction between the incident wave and nuclear potential may lead to partial reflection of the incident wave, called *shape elastic scattering.* The part of the wave function which enters the nucleus undergoes absorption. This process leads to the second stage. Some of the possible absorption processes are :

1. Ejection nucleons in a collisions with incident particle a direct interaction,
2. Multiple collisions of the incoming particle with several nucleons of the target nucleons,
3. The excitation of some type of collective motion such as surface vibrations of the target nucleus,

4. The formation of compound nucleus, without remembering details of the initial stage of formation.

The third stage of the reaction is the more or less rapid transition to the final state. It is concerned with the way the reaction products are produced, *i.e.,* on the second stage. Either disintegration or de-excitation of the compound nucleus will occur. Actually there is no sharp division between the different possibilities in a compound system. Two extreme cases, direct interaction and the compound nucleus, correspond to two general aspects individual and collective particle movements respectively.

STATISTICAL THEORY OF NUCLEAR REACTIONS

The theoretical treatment of the interactions between an incident particle and the nucleons in the nucleus requires the use of a nuclear model. In light nuclei the energy levels of the compound nucleus are generally well separated and the theory of the resonance reaction. In the heavy nuclei, the levels are closely spaced. If the excitation energy is sufficiently high, overlapping levels will be excited and the no. of excited levels is too great to be treated individually, and thus requires Fermi statistics. The compound nucleus is pictured then as a hot liquid or solid from which various particles evaporate. From heavy nuclei, neutrons evaporate most easily even at high energies of excitation but proton emission is also common from lighter nuclei.

In statistical thermodynamics, the entropy S(E) of the system is related to ρ(E), the density of states per unit energy interval, according to the Boltzmann formula as

$$S(E) = \log \rho\,(E). \qquad ...(117)$$

More formal derivation of eqn. (117) is also obtained by defining the thermodynamic free energy F(β) of the nucleus at temperature T = 1/β as

$$e^{-\beta F} \int \rho(E')e^{-\beta E'}\,dE'. \qquad ...(118)$$

We can invert the Laplace transform to obtain

$$\rho(E') = \frac{1}{2\pi i}\int e^{\beta(E'-F)}\,d\beta. \qquad ...(119)$$

This integral can be solved by using *steepest descent method (saddle point method).* The integrated has a saddle point at β = β(E) given by d(βF)/dβ = E, where E is assumed as independent variable. This gives that

$$\frac{d}{dE}[\beta(E-F)] = \frac{d}{dE} = \beta = \frac{1}{T}, \qquad ...(120)$$

where $$S(E) = \beta(E-F) = \int_0^E \beta(E')dE$$

Thus integration of eqn. (119) gives

$$\rho(E) = \frac{e^S}{\sqrt{\left(-\frac{2\pi dE}{d\beta}\right)}}.$$

$$\therefore \quad \frac{d}{dE}\log\rho(E) = \frac{dS}{dE} - \frac{1}{2}\frac{d}{dE}\log\left(-\frac{dE}{d\beta}\right) = \frac{1}{T} - \frac{1}{2}\frac{d}{dE}\log\left(T^2\frac{dE}{dT}\right)$$

As the last term of this eqn. is negligible, hence we get

$$\frac{d}{dE}\log\rho(E) \simeq \frac{d\,S(E)}{dE} = \frac{1}{T}. \qquad ...(121)$$

This is nothing but eqn. (117). Let us now find the dependence of nuclear entropy S and the energy E of the nucleus on the nuclear temperature T. If the nucleons in nucleus are assumed to be as an ideal Fermi gas, we know that the thermal heat capacity $c = dE/dT$ is proportional to T and, therefore, $E \propto T^2$. Hence

$$E = aT^2, \qquad ...(122)$$

where a is a constant. From eqns (122) and (121), we get

$$\frac{d}{dE}\log\rho(E) = \left(\frac{a}{E}\right)^{1/2} \quad \text{or } \rho(E) \sim e^{2(aE)^{1/2}} \qquad ...(123)$$

As approximate result obtained by Ericson is as given below.

$$\rho(E) = \frac{1}{16}\left(\frac{\pi^2}{a}\right)^{1/4} E^{-5/4}\, e^{2(aE)^{1/2}} \qquad ...(124)$$

DIRECT REACTIONS

A direct reaction is one which proceeds without the formation of a compound nucleus. The time during which the incident particle interacts with the target nucleus is very much shorter than the life of a correspond compound nucleus. Formation of compound nucleus is more likely at low energies whereas the direct reaction mechanism will prevail at higher energies. The term direct reaction is used for a variety of nuclear processes including *inelastic nuclear collisions, stripping* and its inverse, the *pick up reaction.* We shall discuss briefly the stripping reaction and its inverse reaction.

Stripping Reactions

The term stripping is used for a type of direct reaction in which the incoming compound particle splits into two fragments, one of which absorbed by the target nucleus and the other continues more or less undisturbed. In this type of reaction target nucleus captures one or two, sometimes three nucleons from the incident particle without the formation of a compound nucleus as an intermediate stage. The remaining portion of the projectile is usually proton, neutron or deuteron. Stripping reactions of several types, *e.g.*, (d, p), (d, n), (t, p), (t, d) and (α, p) are known to occur at high particle energies with many different nuclei. In pick-up reaction incident particle removes one or two nucleons from the target nucleus. Reactions of this type are (p, α), (p, t), (p, He^3). (d, t) and (d, He^3). The theories of stripping and pick-up reactions are very similar and the angular distributions are also quite similar.

Stripping reaction was first recognized by Oppenheimer and Phillips in 1935 in analyzing the low energy (d, p) reactions. It was observed experimentally that (d, p) reactions were more frequent than (d, n) reactions. They explained the reaction by stating that the deuteron behaves as a resistively loose combination of a neutron and a proton. When it approaches a nucleus, the electrostatic repulsion of the positive charges tends to force the proton away, but the neutron is not affected. If the energy of the incident deuteron exceeds the neutron binding energy (2 MeV), the proton will be repelled and neutron will enter the target nucleus. The (d, n) reactions are of particular interest as they are used as a means for obtaining neutrons of high energy. In the high energy region the (d, p) and (d, n) reactions are equally probable.

At medium energies (10 – 20 MeV), the stripping reactions has been very useful for the determination of the energies as well as the spins and parities of excited states of nuclei. In the (d, p), say, stripping reaction the target nucleus accepts a neutron of crbital angular momentum l_n

directly into one of the levels of the final nucleus. The proton proceeds in a direction determined by l_n and with an energy equal to the reaction energy Q for the formation of the level into which its partner was captured plus the kinetic energy of the deuteron. Consider a vector diagram for the (d, p) reaction. The deuteron approaches with momentuın $k_d\hbar$ and the proton goes off with momentum $k_p\hbar$ at an angle θ with the direction of incident deuteron. The momentum of captured neutron may be obtained from the cosine rule applied to momentum triangle.

$$k_n^2 = k_d^2 + k_p^2 - 2k_d k_p \cos\theta \,. \qquad ...(125)$$

A semi-classical consideration of the conservation of angular momentum suffices to give a qualitative understanding of the angular distribution of the protons after the (d, p) reaction. At the point of impact of the deuteron on the surface of the nucleus, there is a splitting into proton and neutron. The neutron subsequently penetrates the nucleus having an impact parameter r and is captured by it. The transfer of momentum $k_n\hbar$ is associated with the transfer of a quantized orbital angular momentum. Hence we have

$$(k_n \times r)\,\hbar = l_n\hbar \text{ or } k_n R \sin\beta = l_n. \qquad ...(126)$$

If $l_n/k_n > R$, the nuclear radius, it is not possible for the d-n interaction to occur. This in order to occur a stripping process, we must have $l_n/k_n \le R$. Hence

$$k_d^2 + k_p^2 - 2k_d k_p \cos\theta \ge \frac{l_n^2}{R^2} \qquad ...(127)$$

or

$$\cos\theta \le \frac{k_d^2 + k_p^2}{2k_d k_p} - \frac{l^2}{2k_p k_d R^2}. \qquad ...(128)$$

The first term on the R.H.S. is always ≥ 1, as $a^2 + b^2 \ge 2ab$. The second term increases quadratically with the quantum number l. There is a lower limit for the angle θ that depends on l_n. Equation (128) shows that the preferred angle of scattering θ will increase with increasing l_n. Choosing

$$E_d = 7 \text{ MeV} \qquad (\therefore k_d = 0.82 \text{ f}^{-1}),$$

$$E_p = 13 \text{ MeV} \qquad (\therefore k_p = 0.79 \text{ f}^{-1}),$$

R = 6f and applying cosine rule we get θ_{max} = 16°, 29° and 42° corresponding to l = 1, 2, 3 respectively. For each l_n values there are secondary maxima also. In general the differential cross-section is largest for $l_n = 0$ and decreases as the angular momentum transfer increases. The position of a peak or a pattern of peaks in the angular distribution of the emitted protons yields the value of l_n.

The analysis of the angular distribution of these protons provides information about angular momentum of these captured neutrons and on the spin and parity of the final nucleus. Let us consider the initial and final states of the nucleus with angular momenta I_i, I_f, the orbital angular momentum of deuteron with respect to the target nucleus l_d and the deuteron spin s_d. If in (d, p) stripping reaction, the proton escapes at an angle θ with respect to the initial direction of motion of the deuteron and neutron is captured, with orbital angular momentum l_n, then we can write

$$I_f = I_i + I_d + s_d - I_p - s_p = I_i + I_n + s_n. \quad ...(129)$$

The orbital angular momentum of the captured neutron l_n is thus restricted by the inequality

$$I_f + I_f + \frac{1}{2} \geq l_n \geq || I_i - I_f \left| - \frac{1}{2} \right| \quad ...(130)$$

Law of conservation of parity $II_f = II_i(-1)^{ln}$ demands that initial and final states have same parity if l_n is even, and opposite parities if l_n is odd.

The quantum theory of low energy stripping reactions was first developed by Butler, using Born approximation which assumes that the plane or any particle whatever close to the nucleus is not distorted by the presence of nuclear potential.

Agreement with experiment is excellent over a wide range of target nuclei and deuteron energies. Improvements have been made which lead to better agreement. They require the use of DWBA (distorted wave Born approximation). There are a number of and it is very likely that a better fit could be obtained.

At very high energies the stripping phenomenon may be described by considering one of the nucleons as absorbed by a nucleus and the other as continuing along its trajectory almost unperturbed with the initial velocity of the deuteron. If proton is captured by the target nucleus, the energy spread and the angular spread of the neutron beam are

$$\Delta E = 1.5\,(E_d B_d)^{1/2}, \text{ and } \Delta^{\theta} = 1.6 \left(\frac{B_d}{E_d}\right)^{1/2} \quad ...(131)$$

where B_d is the binding energy of the deuteron and E_d its kinetic energy.

NEUTRON COLLIMATORS

Neutrons produced by the above sources possess very high energy range. Most of the neutrons are called fast neutrons as they have enormous velocities. Since the ordinary methods used for charged particles and γ-rays can not be used for the collimating a beam of fast neutrons. Dunning (1938) has devised a collimator known as the *neutron howitzer*. He used the property of slowing down of neutron in hydrogeneous materials like paraffin or water. The radium beryllium source S is placed at the bottom of a barrel inside a paraffin block. A cylindrical slit is cut in this block which is lined from all sides by cadmium sheet. Neutrons are emitted in all directions. The neutrons energy is negligibly small in all directions except through the barrel. The thermal neutrons resulting from the slowing down of the fast neutrons in the paraffin are effectively absorbed by the cadmium coating over the outer surface a paraffin and walls of the barrel. The cadmium layer is covered by a layer of lead to g-rays arising from neutron capture.

NEUTRON SPECTROSCOPY

Thick photographic emulsion is the simplest example of a neutron spectrometer. Neutrons passing through the emulsion may collide with protons in it and give up a fraction or practically all of their kinetic energy. The energy of the proton is determined by measuring the length of the track, and the neutron energy can then be computed as $T_n = T_p (\cos\theta)^{-1}$. Mere θ is the angle between the directions of the proton and the incident neutron. The direction of the entering neutron is presumably known and the direction of the proton or the track in the emulsion is measured under the microscope. The resolving power is limited by the straggling of the protons in the emulsion.

Scintillators containing hydrogen can be used to detect the protons recoiling from n-p collisions in the scintillator material. Unless the scintillator is very large, the total energy of the neutron is not given to the scintillator. Since it is very difficult to measure the angle between the neutron direction and the proton direction, hence the scintillation counter is not very well suited for neutron spectroscopy. Solid state detectors have recently come into use in neutron spectroscopy in the so-called *sandwich counters*.

A thin film containing Li^6 is sandwiched between two solid state counters. In passing through the sandwich neutrons (1 to 10 MeV) will have a reasonably high probability of initiating the Li^6 (n, t) He^4. If the alpha particle enters one of the solid state counters and the triton enters the other, a coincidence can be registered and the sum of the pulse heights from these counters is a measure of the neutron energy. The detection efficiency of sandwich counters is of the order of 10^{-6} or 10^{-7}.

With pulsed accelerators and devices to measure the time of flight of neutrons over a path from target to detector, the most exact work in neutron spectroscopy has been performed. The energies are determined through a measurement of velocities of neutrons.

The accelerator may be a Van de Graaff generator and the detector may be a scintillation counter with associated electronic equipment designed specially 10 measure the time delay (the time in which a pulse strikes the target) spectrum. The flight time for neutrons over a path length L in metres depends on the kinetic energy T (in MeV) of the neutron as

$$t = \frac{L}{v} = L\sqrt{\frac{m}{2T}} = \frac{L}{c}\sqrt{\frac{mc^2}{2T}} = \frac{72.3L}{\sqrt{T}} \text{ n sec.} \qquad \text{...(16)}$$

This is quite good for neutron spectroscopy but decreases with the increase of neutron energy because time of flight decreases with the K.E. of the neutron while Δt remains constant. The resolving power can be increased by increasing the time of Eight by making the path longer. The time of flight spectrum of neutrons from a Ne^{20} gas target bombarded with 3.07 MeV deuterons.

The intense peak to the far right corresponds to the γ-rays arriving first (v = c). The target to counter distance L was 2.02 m. The time difference between channels was approximately 2.3 nsec with zero time at about channels 102.

CHARGED PARTICLE REACTION SPECTROSCOPY

The scintillation counter has proved very much useful when too high resolving power is not required and is specially used when coincidences are being studied between the charged particles and γ-rays de-exciting the state formed in the reaction. The scintillator can be mounted quite close to the target to yield a large sòlid angle of acceptance and to help to cut down the background pulses from γ-rays, neutrons and cosmic rays.

It should be very thin because the charged particles in low energy nuclear spectroscopy have a relatively small range in solid materials. The scintillation spectrometers attain the resolving power $E/\Delta E = 20$ to 30.

For high resolution work, the magnetic spectroscopes are used in nuclear reaction studies. Single focusing instruments, as used in α-spectroscopy, produce a line image of a point source located in the median plane of the magnet, at some distance from the exit of the magnet $\perp$ to the median plane because there are no focusing forces acting perpendicular to the median plane. Point image of a point source is produced by double focusing instruments, can be designed in various ways. A very popular spectrometer of this kind utilizes a magnet with an in homogeneous field similar to the instrument used β-spectroscopy except that it is a sector magnet with both the source and the detector outside the field.

We shall describe now a complex instrument used in nuclear reaction studies in association with an 8 MeV Van de Graaff machine, in more detail. Actually in this spectrograph twenty four instruments are placed in one large vacuum enclosure. Each spectograph records a broad range spectrum of particles emitted at a given angle θ from the incident beam. In steps of 7.5° the range of angles is 0° to 172.5°. The toroidal magnetic circuit comprises forty eight 7.5° cast iron wedges, some of which are thinner than others so that gaps are left between them. These are rigidly clamped together with stainless-steel spacers defining the gaps. Twenty five of the gaps are 3/8" wide, and one gap is 1.5" wide. Each of the former gaps is magnetized by five turns on each of the two pole pieces adjacent to the gap.

The latter gap is magnetized by twenty turns on each side of the gap. The beam reaches the target (placed in the centre of the torus) region by passing through a field free hole through the iron. Lower part of the spectrograph with the beam entering hole, the target and the particle trajectories in the 0° gap. The pole face geometry and focal surface geometry are identical for all gaps. The range of energy covered in one exposure is approximately 2.3 : 1, *i.e.,* if 11.5 MeV protons are focussed to the top end of the focal surface, 5 MeV protons will he focussed to the lower end.

After passing through the energy defining slit of the beam analyzing magnet, the beam passes through an electrostatic quadrupole lens designed to produce a reduced image of the energy defining slit on the target. This lens is situated inside a low-carbon steel shielding tube in the 3" diameter hole in the iron torus. It has magnification 1/3 in the vertical direction and unity in the horizontal direction. The target chamber consists of a low-carbon steel pipe of 24" diameter, closed at the bottom by a stainless steel plate used as a reference surface and at the top by a liquid nitrogen cooled copper plate. The beam, after passing through the target, can be collected by a Faraday cup at the opposite end of the target chamber.

The reaction products are recorded on nuclear track plates mounted in rigid holders above each gap. The plates, after exposure, are developed and scanned under microscopes in half millimeter strips across the plate. The number of tracks having correct length and direction are counted. The number of 1/2 mm strips in 30" is about 1500. This quantity when multiplied by the number of angles normally utilized (24) gives 36,000 data points obtainable with this instrument in one single exposure. The over all resolving power of such an instrument is not determined by the instrument alone but, in general, by the following factors:

(a) the energy stability of the beam of the Van de Graaff generator,
(b) the vertical dimension of the beam spot on the target,
(c) the width of the strips scanned with the microscope across the nuclear track plates,
(d) the effective target thickness and
(e) the aberration of the spectrograph. The resolving power of the above described apparatus was $E/\Delta E \sim 10^3$.

CONSERVATION LAWS

We shall merely list the various conservation laws that appear to be valid in ordinary nuclear interactions.

1. Conservation of Angular Momentum

The total angular momentum *l* comprising the vector sum of the intrinsic angular momentum s and relative orbital momentum *l* of the products must be equal to the total angular momentum of the initial particles.

2. Conservation of Energy

The total energy of the products, including both mass energy and kinetic energy of the particles plus the energy involved must be equal to the mass energy of the initial ingredients plus the kinetic energy of the bombarding particle.

3. Conservation of Momentum

The total linear momentum of the products must be equal to the linear momentum of the bombarding particle (the target nucleus is ordinarily taken to be at rest).

4. Conservation of Isotopic Spin

The invariance of the nuclear Hamiltonian function towards the charge character of the nucleons can be expressed analytically as an invariance towards rotational shifts of the axes in isotopic spin space, and there should correspondingly exist a conservation law for the isotopic spin of a nuclear system.

5. Conservation of Parity

The parity of the system determined by the target nucleons and bombarding particle must be conserved throughout the reaction. The total parity of the system is the product of intrinsic-parities of the target nucleus and bombarding nucleus. No violation of parity has been observed in a nuclear reaction (strong nuclear forces). Although parity does not appear to be conserved in weak interactions.

6. Conservation of Spin

The important conservation law asserts that the spin character or a closed system cannot change, *i.e.,* the statistics remains same as that existed before reaction.

7. Conservation of Nucleons

The law of conservation of nucleons states that the nucleons can neither be created nor can be destroyed so that the number of nucleons minus the number of anti-nucleons in the universe remains constant.

8. Conservation of Charge

The total electric charge of the products must be equal to the total electric charge of initial particles.

NEUTRONS DISCOVERY

Rutherford in his Bakerian Lecture to the Royal Society in 1920 said, *"Under some conditions, ..if may be possible for an electron to combine much more closely with the hydrogen nucleus (i.e., a proton), forming a kind of neutral doublet. Such an atom would have novel properties. Its external field would be practically zero...and consequently it should be able to move freely through matter. Its presence would probably he difficult to detect.*" In 1930, Bothe and Becker bombarded lithium, beryllium and boron with α-particles from polonium and found a very penetrating but nonionizing radiation, they assumed that the radiation was of the γ-ray type because of its tremendous penetration. While repeating these experiments in 1932, Dr. H. Joilot and his wife, Dr. Irene Curie Joliot, daughter of Madam Curie found that which a sheet of hydrogen containing material, particularly paraffin, was interposed in the path of these radiations, protons were ejected with a considerable velocity. From the ranges of these recoil protons, the maximum proton energy E proved to be about 5.3 MeV. Assuming that the protons were produced as the result of elastic collisions with the *γ-ray photons*, calculations showed that each photon must have possessed an amount of energy of about 52 MeV. These results were entirely inconsistent with the results from the experiments on the absorption of these rays in lead (about 7 MeV). For an α-particle of 5 MeV energy :

$$_4Be^2 + _2He^4 + E_a \rightarrow (_6C^{13}) \rightarrow \; _6C^{12} + h\nu$$

$$9.01503 + 4.00388 + .00536 = 13.00751 + h\nu$$

or $$h\nu = 0.01676 \text{ amu}, = 15.5 \text{ MeV}.$$

This is much shorter than the required energy in order to explain the proton scattering. At Cambridge, Webster observed that the energy of the photons emitted by the beryllium varied according to their direction of emission, heaving least value if emitted in the direction opposite to that of the incident α-particles. The situation was resolved by Chadwick, in England in 1932, after performing a series of measurements of the energies of recoil of protons ejected from thin targets by the penetrating "Be-radiation" with a pulse ionizing chamber and amplifier. He said, "*These results are very difficult to explain on the assumption that the radiation from beryllium is a quantum radiation, if energy and momentum are to be conserved in the collisions. The difficulties disappear, however, if it be assumed the radiation consists of particles of mass* 1 *and charge 0, or neutrons.*" Chadwick's theory made sense of Webster's puzzling discovery that the neutron energy varied with the direction of emission. These neutrons were formed as a result of the highly exoergic nuclear reaction

$$_4Be^9 + _2He^4 \rightarrow (_6C^{13}) \rightarrow \; _6C^{12} + _0n^1 + 5.65 \text{ MeV}.$$

Thus we see that *comparatively late date of the discovery of the neutron is partly due to the similarity in the properties of the neutron and high energy photons and the confusing effect that γ-rays as well as neutrons are associated usually with neutron sources.*

NEUTRON SOURCES

Photo-neutron Sources

When γ-rays of sufficient energy (greater than the neutron binding energy) are incident upon light elements a probability exists that neutrons will be emitted by the energized nuclei. The majority of nuclei have a neutron binding energy of 5 MeV or more, except $_4Be^9$ (1.67 MeV) and $_1H^2$ (2.23 MeV). g-rays from artificially produced radio-nuclei can serve in photo disintegration sources.

A standard source consisting of a *MsTh* or Na^{24} gamma source is placed at the centre of a spherical vessel filled with heavy water as shown in Fig. 4.1. Yield, the number of neutrons per sec due to 1 gm. of Be or D_2O at a distance of 1 cm. from 1 curie, of various gamma sources is given in the table 1.

Table 1 : Radioactive (γ, n) sources.

γ-Source	Target	Half Life	E_γ (MeV)	E_n (MeV)	Yield
Na^{24}	Be	14.8 hr.	2.76	0.83	13×10^4
Na^{24}	D_2O	14.8 hr.	2.76	0.22	27×10^4
Y^{88}	Be	87 d.	1.9, 2.8	0.16	10×10^4
Y^{88}	D_2O	87 d.	2.8	0.31	0.3×10^4
Sb^{12}	Be	60 d.	1.7	0.025	19×10^{24}

The advantages of this type of source are :

(a) The reproducible neutron strength,

(b) The choice of E_n by choosing a suitable value of E_γ.

(c) The wide variety of artificial γ-sources available from atomic reactors,

(d) The neutrons produced are more mono-energetic than those obtained from (Be – α) sources.

Wattenberg gave the equation for the energy E_γ of the emitted neutrons as

$$E_n = \frac{A-1}{A}\left[E_\gamma - Q - \frac{E_\gamma^2}{1862(A-1)}\right] + E_\gamma \cos\theta \left[\frac{2(A-1)(E_\gamma - Q)}{931\,A^3}\right]^{1/2}$$

where A is the mass of the target nucleus, E_γ the energy of γ-ray in MeV, the threshold energy in MeV for the (γ, n) reaction in the target nucleus and θ is the angle between the path of the γ-ray and the direction of the emitted neutron. As Na^{24}-source is of short half life and hence Sb^{124} is very convenient for intermediate energy neutrons with a relatively long half life.

Radioactive Sources

The majority of the neutrons is produced by the (a, n) reaction given above. If the radium is in secular equilibrium with its decay products, extra deuterons are released according to the reaction

$$_4Be^9 + h\nu \rightarrow {}_4Be^8 + {}_0n^1.$$

The product nucleus of this reaction Be^8, is unstable and thus decays into two a-particles within 10^{-15} sec. This neutron source is isotropic and produces neutrons over a wide energy range (0 → 13 MeV). The neutrons are not mono-energetic due to following reasons :

(a) The α-particles do not all have the same initial energy.

(b) Many α-particles lose part of their initial energy by ionization before nuclear capture,

(c) The neutron energy varies with the direction of emission having least value if emitted in the direction opposite to that of the incident α-particles.

(d) $_6C^{12}$ can be left in an excited state.

We can have slow neutrons after putting the Ra – α – Be source in water, paraffin, or some other hydrogen-containing material. For a 2-MeV neutron only 25 elastic collisions with protons are needed. Radon is some times used instead of radium, as it is a gas and can be made into a more compact source. But it has the disadvantage of decaying rapidly because of the snort half life of 3.8 days. Polonium is used instead of radium when a neutron source with relatively few g-rays is desired. This has several disadvantages :

(i) the neutron output is much lower (3 × 10^6 neutrons/sec per curie of Po while it is 10 – 18 × 10^6 neutrons/sec per curie of Ra for radium),

(ii) special chemical facilities are needed to prepare Po,

(iii) the half life is short (140 days) — compared with that of Ra – Be source (1600 years).

Reactor Sources

The most powerful source of neutrons available in these days is a nuclear reactor or chain reacting pile on which uranium undergoes fission. The fission of each heavy nucleus by slow neutron capture is accompanied by the release on the average of about 2.5 neutrons. The neutrons produced from the fission on of uranium atoms are fast and are slowed down in the moderator. The disadvantages of reactor sources are the wide spread in neutron energy and the impossibility of directly pulsing the neutron source.

X-rays Sources

Through the (γ, n) reaction, high energy X-ray machines can produce the high energy neutrons. A flux of high energy photons is produced when the electrons accelerated to high energies in linear accelerators, betatrons, synchrotrons etc. impinge on target elements of high atomic number. A large number of neutrons within the target material are released by the high flux of photons. The disadvantage of this type of source is the wide spread in neutron energy.

Ultrafast Neutrons

As a deuteron behaves as a relatively loosely bound system consisting of a neutron and a proton which are frequently outside the range of their mutual forces, hence, deuterons accelerated to about 200 MeV in a synchro-cyclotron can readily be "stripped" into protons and neutrons by directing the deuteron beam to pass through a target nucleus. Recently neutrons of energy > 300 MeV have been produced by stripping 650 MeV deuterons accelerated in the Birmingham synchrotron. Almost any element may be used as the target material for stripping out the yield varies with its nature.

Neutrons from Accelerated Particle Reactions

Neutrons are produced by (d, n) and (p, n) reactions in particle accelerators. These sources give mono-energetic neutrons in the range from a few keV to 20 MeV. In deuteron bombardment the energy of the released neutron is progressively larger for most of the target elements, which is due to the low binding energy of the deuteron. The C^{12} (d, n) N^{13} reaction is an exception to the rule with a Q value of – 0.28 MeV. Those neutrons which are emitted in the direction of the incident beam of deuterons will have a maximum energy, equal to the sum of the incident and the reaction energies approximately. Some (d, n) reactions are:

$${}_1H^1 + d \rightarrow {}_2He^3 + {}_0n^1 + 3.265 \text{ MeV}$$

$${}_1H^3 + d \rightarrow {}_2He^4 + {}_0n^1 + 17.6 \text{ MeV}$$

$${}_3Li^7 + d \rightarrow {}_4Be^8 + {}_0n^1 + 15.0 \text{ MeV}$$

$${}_4Be^9 + d \rightarrow {}_5B^{10} + {}_0n^1 + 3.79 \text{ MeV}.$$

Several light and moderately heavy elements undergo (p, n) reactions. These endoergic reactions are the convenient sources of neutrons of moderate energy. Some (p, n) reactions are :

$${}_1H^3 + p \rightarrow {}_2He^3 + {}_0n^1 - 0.764 \text{ MeV}$$

$${}_3Li^7 + p \rightarrow {}_4Be^7 + {}_0n^1 - 1.646 \text{ MeV}.$$

The Li^7 (p, n) Be^7 reaction suggests that accelerated charged particles can be used to produce nearly monoenergetic neutrons.

SOLVED EXAMPLES

Example 1:

A tritium gas target is bombarded with a beam of monoenergetic protons of kinetic energy 3 MeV. What is the K.E. of the neutrons emitted at 30° to the incident beam? Atomic masses are : H^1 1.007276 mu, n^1 = 1.008665 mu, ${}_1H^3$ = 3.016056 mu, and He^2 = 3.016030 mu.

Solution:

Nuclear reaction in question can be written as

$${}_1p^1 + {}_1H^3 \rightarrow {}_2He^3 + {}_0n^1 + Q,$$

where $Q = m({}_1p^1) + m({}_1H^3) - m({}_2He^3) - m({}_0n^1)$

= 1.007276 + 3.016050 – 3.016030 – 1.008665

= – 0.001369 mu

= – 1.2745 MeV.

The K.E. associated with the emitted neutron is given by eqn. (11) as

$$\sqrt{E_n} = u \pm \left(u^2 + \nu\right)^{1/2},$$

$$\text{where } u = \frac{\left(m_p m_n E_p\right)^{1/2}}{mH_e + m_n}\cos\theta = \frac{(1.007276\times1.008665\times3)^{1/2}}{3.016030+1.008665}\times\frac{\sqrt{3}}{2}$$

$$= 0.3753$$

$$\text{and } \nu = \frac{m_{He}Q + E_p\left(m_{He} - m_p\right)}{m_{He} + m_n}$$

$$= \frac{-3.016030\times1.2745 + 3\times2.008754}{3.016030+1.008665} = 0.5424$$

$$\therefore \sqrt{E_n} = 0.3753 \pm 0.8266 = 1.2019$$

$$\text{or } E_n = 1.444 \text{ MeV.}$$

Example 2:

When F^{19} is bombarded with protons a (p, n) reaction with subsequent α-emission occurs. Calculate the excitation energy of the compound nucleus that corresponds to the resonance with a proton energy of 4.99 MeV.

$${}_1H^1 + {}_9F^{19} \rightarrow ({}_{10}Ne^{10})^* \rightarrow {}_{10}Ne^{19} + {}_0n^1.$$

The convertible energy of the protons or the energy of relative motion is

$$E = \frac{E_0}{1+\frac{m}{M}} = \frac{4.99}{1+\frac{1}{19}} = 4.745 \text{ MeV.}$$

This binding energy contribution of the absorbed proton in the compound nucleus is

$$\Delta E = [M({}_9F^{19}) + M({}_1H^1) - M({}_{10}Ne^{28})]mu$$

$$= [18.998405 + 1.007276 - 19.992440]mu$$

$$= 0.013241 \text{ mu} = 12.332 \text{ MeV}$$

∴ Excitation energy of the compound nucleus

$$= 4.745 + 12.332 = 17.077 \text{ MeV}$$

Example 3:

A thin sheet of Co^{59}, 0.04 cm, thick, is irradiated with a neutron beam of flux density 10^{12} neutrons per cm^2 per sec for a period of 3 hr. If the cross-section for neutron capture by Co^{59} is 30 barns, calculate the number of nuclei of the isotope Co^{50} produced at the end of the irradiation period per cm^2 and the initial β-activity of the sample. Given: half-life of Co^{60} 5.2 years and density of Co^{59} 8.9 gm/cm³.

Solution:

$$\text{Number of reactions/second/m}^2 = N_0\sigma It = N\left(\frac{\rho}{M}\right)\sigma It$$

$$= 6.02252 \times 10^{26} \times \left(8.9 \times \frac{10^3}{59}\right) \times 30 \times 10^{-28} \times 10^{16} \times 0.04 \times 10^{-2}$$

$$= 0.092 \times 10^{15}.$$

$\therefore$ No. of Co^{60} nuclei produced = No. of transmutation in 3 hours

$$= 1.092 \times 10^{15} \times 3 \times 60 \times 60 = 118 \times 10^{17}.$$

$$\text{Initial activity per m}^2 = \lambda \times 118 \times 10^{17} = \frac{0.693 \times 118 \times 10^{17}}{5.2 \times 365 \times 24 \times 60 \times 60}$$

$$= 5 \times 10^6 \text{ decays/sec} = 134 \text{ micro curies..}$$

Example 4:

Estimate the relative probabilities of (n, n) and (n, γ) in indium, known to have a neutron resonance at 1.44 eV with a Γ of 0.1 eV and cross-section of 28000 barns.

Solution:

At resonance ($E = E_0$) Breit-Wigner formula simplifies to

$$\sigma(n, \gamma) = \frac{\lambda^2 \Gamma_n \Gamma_\gamma}{4\Gamma^1}.$$

The de Brogile wavelength $\lambda = \dfrac{h}{\sqrt{(2mE)}} = 2.4 \times 10^{-11}$ m.

$$\Gamma_n \Gamma_\gamma = \frac{4\sigma\Gamma^2}{\lambda^2} = \frac{4 \times 2.8 \times 10^{-24} \times (0.1)^2}{(2.4 \times 10^{-11})^2}$$

$$= 1.5 \times 10^{-4}$$

Since $\Gamma_\gamma \simeq \Gamma = 0.1$, hence $\dfrac{\Gamma_n}{\Gamma_\gamma} = 0.015.$

Example 5:

The neutron capture reaction of Au^{197} at neutron energies upto a few hundred eV is characterized by a number of resonances. The most prominent is at 4.906 eV and has $\Gamma_\gamma = 0.124$ eV and $\Gamma_n = 0.007\ E^{1/2}$ eV. The compound nucleus formed in this resonance absorption has spin 2. Calculate the peak cross-section.

Solution:

$\Gamma_n = 0.007\ E^{1/2} = 0.007 \times (4.906)^{1/2} = 0.0155$ eV.

We know the relation for resonance cross-section

$$\sigma(n,\gamma) = \frac{\lambda^2}{\pi}\frac{2I_c+1}{2(2I+1)}\frac{\Gamma_\gamma\Gamma_n}{\left(\Gamma_\gamma+\Gamma_n\right)^2} = \frac{h^2}{2\pi mE}\frac{2I_c+1}{2(2I+1)}\frac{\Gamma_\gamma\Gamma_n}{\left(\Gamma_\gamma+\Gamma_n\right)}$$

$$= \frac{\left(6.625\times10^{-34}\right)^2}{2\times3.14\times1.67\times10^{-27}\times4.906\times1.6\times10^{-15}} \times \frac{2(2+I)}{2(3+1)} \times \frac{0.124\times0.0155}{(0.1395)^2}$$

$$= 3.446 \times 10^4 \text{ barns.}$$

Example 6:

Estimate by the aid of the semi-classical stripping theory the angles at which the (d, p) differential cross-section has its maximum for the following l-values : l = 0, 1, 2, 3, 4. Use R = 5f, E_d = 10 MeV and E = 14 MeV.

Solution:

From semi-classical theory for stripping reaction, the angles are given by the reaction

$$\cos\theta \le \frac{\left(k_d^2+k_p^2\right)}{2k_d^2} - \frac{l^2}{2k_d k_p R^2},$$

where $k_d = \sqrt{\frac{(2m_d E)}{\hbar}} = 0.98\ f^{-1}$ and $k_p = \sqrt{\frac{(2m_p E)}{\hbar}} = 0.82\ f^{-1}$.

$$\therefore \qquad \cos\theta \le \frac{0.98^2 + 0.82^2}{2\times0.98\times0.82} - \frac{l^2}{2\times0.98\times0.82\times5^2}$$

$$\le 1.015 - \frac{l^2}{40.18}.$$

Hence from natural cosines table, we get

l	0	1	2	3	4
θ_{max}	0	90°	24°	38°	52°

Example 7:

Calculate the Q-value for the formation of P^{30} in the ground state in the reaction Si^{29} (d, n) P^{30} from the following cycle of nuclear reactions:

(i) $P^{31} + \gamma \rightarrow P^{30} + n - 12.37$ *MeV*

(ii) $P^{31} + p \rightarrow Si^{28} + He^4 + 1.909$ *MeV*

(iii) $Si^{28} + d \rightarrow Si^{29} + p + 6.246$ *MeV*

(iv) $2d \rightarrow He^4 + 23.834$ *MeV*

Solution:

Reaction (i) can be written as

$P^{30} + n \rightarrow P^{31} + g + 12.37$ MeV. ...(ii)

Formation of P^{30} is given by the reaction

$Si^{29} + d \rightarrow P^{30} + n + Q.$

Combination of these equations with reactions (ii), (iii) and (iv) gives

$4d = 2He^4 + Q + 44.359$ MeV

$Q = 2(2d - He^4) - 44.359$

$= 47.668 - 44.359 = 3.309$ MeV.

EXERCISES

1. The B^{10} (α, p) C^{13} reaction shows among other a reasonance for an excitation energy of the compound nucleus of 13.23 MeV.

Calculate the mean life of the nucleus for this excitation if level width is 130 keV.

2. Calculate the threshold energy required to initiate and the reaction P^{31} (n, p) Si^{31}, given M_p = 1.00814, M_n = 1.00898, M_P = 30,98356 & M_{Si} = 30.98515.

   Calculate also the maximum energy of β-decay of Si^{31} to P^{31}.

3. In the B^{10} (α, p) C^{13} reaction with 4.77 MeV α-particles the two most energetic proton groups which were observed to be emitted at the angle of 90° with the direction of the incident α-particle beam had energies of 6.84 MeV and 3.98 MeV respectively. Give the information about the energy levels of the residual nucleus.

4. Assuming a resonance cross-section 5.5×10^4 barns for Cd^{113} for Cd^{113} at a resonance energy of 0.176 eV. Calculate the relative probability of a neutron emission as compare to a resonance capture if Γ = 0.113 eV and $\Gamma_\gamma << \Gamma_n$.

5. Calculate and compare the collision and absorption mean free paths for neutrons in graphite. Given : σ_s = 4.8 barns, σ_a = 3.2 $\times 10^{-3}$ barns and ρ = 2.25 gm/cm^3.

6. An indium foil of 2 cm^2 cross-section and 10^{-3} cm, thickness is exposed to a spread beam of neutrons of uniform energy. Calculate the number of neutron capture that will occur during a 3 min exposure of the foil. Given; I = 5 × 10 neutrons/cm^2 sec. and σ_u = 190 barns.

2

Nuclear Reactors Element

INTRODUCTION

The neutron balance in a thermal reactor can be described in terms of a cycle that shows what happens to the neutrons. Consider this cycle starting with the fission of U^{235} nucleus by a thermal neutron, v fast neutrons are emitted in this process. Some of them, having energy greater than the fission threshold for U^{238}, can cause fissions in U^{238} with a consequent increase in the number of fast neutrons. The total number of fast neutrons from fission is thus increased from v to vϵ, where ε is greater than one and is called the *fast fission factor*. In certain reactors $\epsilon = 1.03$. The particular processes with which we are concerned are nuclear fission reactions in which fresh neutrons are supplied by the fission processes. Thus to maintain a chain reaction, the neutrons must be produced at a constant rate and number of neutrons released per fission, v, must be sufficiently greater than 1 to compensate neutron loss. Since v is constant for a given fissionable material and, therefore, beyond human control.

The only alternative is to reduce the neutron loss. The achievement of a chain reaction with uranium depends on a favourable balance among following six processes:

(1) Neutron may escape without being captured.

(2) Neutron may be absorbed by other materials present in the assembly without causing fission.

(3) Neutron may be absorbed by U^{235} without causing fission.

(4) Neutron may be absorbed by U^{235} nucleus causing fission.

(5) Neutron may be absorbed by U^{238} without producing fission.

(6) Neutron may be absorbed by U^{238} and may cause a fission when its energy is greater than the threshold energy.

Thus we see that processes (1) and (3) create new neutrons while other remaining processes remove available neutrons from the assembly. *A device for releasing nuclear energy through fission at a controlled rate is called nuclear reactor or chain reacting pile.* In nuclear weapons, uncontrolled nuclear reactions are used as most rapid and largest energy released is desired. From a finite size reactor core, some fast neutrons may leak out before being slowed down to thermal energies. To account for this, we introduced l_f the leakage factor for fast neutrons and find that the number of fast neutrons remaining in the core long enough for slow down is $\nu\varepsilon\,(1 - l_f)$. These neutrons are slowed-down by collisions with moderator atoms, but during the slowing down process some of them may be captured by U^{238} to form U^{239} which decays to Np^{239} and then to Pu^{239}, particularly at energies in the region of strong resonances. We introduce the resonance escape probability p and find that out of $\nu\in(1 - l_f)$ neutions, a fraction $\nu\varepsilon\,(1 - l_f)\,p$ escapes while the fraction $\nu\in(1 - l_f)(1 - p)$ is captured and goes to form Pu^{239}. The neutrons which escape are slowed down to thermal energies. Some of these thermal neutrons may leak out of the core. To account for this we introduce l_{th}, the *leakage factor for thermal neutrons* and find that the number of slow or thermal neutrons remaining in the core is $\nu\varepsilon\,(1 - l_f)\,p\,(1 - l_{th})$. A fraction of this remaining number is absorbed by uranium and the remaining other part by other materials. The fraction of thermal neutrons absorbed by the fuel as compared to all thermal neutron absorptions in the assembly is called the *thermal utilization factor f*. If we introduce this term we obtain the number of thermal neutrons that are actually absorbed by the fuel, which is $\nu\in(1 - l_f)\,p\,(1 - l_{th})f$.

Not all of the neutrons absorbed in the fuel caue of U^{238} nuclei, some are absorbed by U^{238} to form Pu^{239} while others are absorbed m U^{235} to form U^{236}. If σ_a is the cross-section of thermal neutrons absorbed in uranium and σ_f that of neutrons cause fission of U^{235}. Thus the final number of second generation neutrons absorbed in the fuel is $\nu\in(1 - l_f)\,p\,(1 - l_{th})\,f\sigma_f/\sigma_a$. This quantity is known as *reproduction factor or multiplication factor* and is denoted by k. The quantity $\nu\,(\sigma_f/\sigma_a)$ represents the number of fast fission neutrons, produced by each thermal neutrons, absorbed by the fuel and is called η. This quantity can be expressed as

$$\eta = \nu\frac{N_{235}\,\sigma_f(U^{235})}{N_{235}\,\sigma_f(U^{235}) + N_{235}\,\sigma_r(U^{235}) + N_{238}\,\sigma_o(U^{238})} \qquad ...(1)$$

where σ_r (U^{235}) stands for the cross-section for irradiative capture by U^{235}. For natural uranium N_{235}/N_{238} = 0.00715. It has been found that v = 2.5. The experimental values for the microscopic cross-section are σ_f (U^{235}) = 549 barns, σ_r (U^{235}) barns, σ_a (U^{238}) = 2.8 barns, hence η = 1.32.

$$\therefore \qquad k\ \eta \in (1 = l_f)\ p\ (1 - l_{th})f \qquad ...(2)$$

If the core is of very large size, the leakage of fast as well as slow neutrons can be disregarded. The multiplication factor is then denoted by kw and is given by

$$k_\infty = \eta\varepsilon pf \qquad ...(3)$$

This relation is known as the *four-factor formula.*

In an infinite system the condition of the self-sustaining chain reaction is that each neutron generation just replaces the previous one *i.e.*, k = 1. The system is then said to be critical. Fora finite system, because in each neutron generation some neutrons are loset from the sides of the system, the *critical condition* is that k must be greater than unity, the exact amount by which it has to be greater than unity depending upon the shape of the system and the arrangement of the materials in the system, etc. If k is greater than unity for the infinite system, the number of neutrons will increase steadily. This is known as a *super-critical* state. On the other hand for k less than unity the number of neutrons would decrease to zero and this state is known as *sub-critical* state.

In the special case of a reactor in which the fuel contains U^{236} only and no U^{238}, both the factors e and p are practically unity. The formula for kw thus reduces to

$$k_\infty = \eta f \qquad ...(4)$$

For an assembly of finite dimensions, the effective multiplication factor keff will be less than k_∞ by a factor $(1 - l_f)\ (1 - l_{th})$. Thus

$$k_{eff} = k_\infty (1 - l_f)\ (1 - l_{th}) \qquad ...(5)$$

The magnitude of keff determines the speed with which the number of neutrons builds up and the rate at which fission occurs. In a nuclear bomb type of assembly, this build up must take place very rapidly, ($k_{eff} > 1$), whereas in industrial and research reactors this self-multiplication must be slow enough to allow the fission rate to remain always under the control of the operator ($k_{eff} \simeq 1$).

If $k_{ef} > 1$ for an assembly, we can decrease it by progressively reducing the reactor size, there by increasing the neutron loss through leakage from the assembly. The reactor size of the assembly for which $k_{eff} = 1$ is called the critical size.

REACTORS CRITICAL SIZE

The size of a reactor that operates under the condition of exact balance between neutron production and neutron loss is known as critical *size and the reactor itself is said to be critical.* For such a reactor, a relation between its geometric properties and the material properties of the assembly is known as the critical equation. For all types of reactors, both homogeneous and heterogeneous, we start with eqn.

$$\nabla^2 \phi - \frac{3}{\lambda_{tr}\lambda_a}\phi + \frac{3q}{\lambda_{tr}} = 0 \qquad ...(6)$$

where q is the slowing down density which is a function of the space co-ordinates r and the neutron age τ. Thus the Fermi age equation can be written as

$$\nabla^2 q(r,\tau) = \frac{\partial q(r,\tau)}{\partial \tau} \qquad ...(7)$$

The solution of this equation can be written in the form of a product of two functions R (r) and T (τ). We can thus write

$$q(r, \tau) = R(r)\ T(\tau). \qquad ...(8)$$

By taking the second derivatives with respect to the space coordinates and first derivative with respect to τ only and substituting in equation (7), we get

$$T(\tau)\ \nabla^2 R = R(r)\ \partial T/\partial \tau$$

or

$$\frac{\nabla^2 R}{R(r)} = \frac{1}{T(\tau)}\frac{\partial T}{\partial \tau} = -B^2 \text{ (say)} \qquad ...(9)$$

where $-B_2$ is a constant. Thus we have two differential equations.

$$\frac{1}{T(\tau)}\frac{\partial T}{\partial \tau} = -B^2 \quad \text{or} \quad \frac{\partial T}{\partial \tau} + B^2\ T(\tau) = 0 \qquad ...(10)$$

and

$$\frac{\nabla^2 R}{R(r)} = -B^2 \quad \text{or} \quad \nabla^2 R + B^2 R(r) = 0 \qquad ...(11)$$

The solution of eqn (10) is

$$T(\tau) = T_0 e^{-B^2 r} \quad \text{...(12)}$$

where T_0 is the value of T initially when $\tau = 0$.

At the beginning of the slowing down process,

$$q(\tau = 0) = q_0 = R(r)\, T(0) = R(r)\, T_0 \quad \text{...(13)}$$

$$\therefore \quad q = R(r)\, T_0 e^{B2} = q_0 e^{-B^2 r} \quad \text{...(14)}$$

We can also express q_0 in terms of the physical properties of the assembly. Number of neutrons per unit volume per sec. available for slowing down is (given by the rate of production of fission neutrons and is equal to the neutron multiplication factor $\in f\eta$ per thermal neutron absorbed. The rate of thermal neutron absorption per unit volume of the reactor is $\phi\Sigma_a$. Hence

Initial slowing down density = rate of fast neutrons available

or $$q_0 = \phi(r)\, \Sigma_a \varepsilon f \eta \quad \text{...(15)}$$

Thus equation (14) becomes

$$q = \phi(r)\, \Sigma_a \varepsilon f \eta\, e^{-B^2 r} \quad \text{...(16)}$$

If we consider neutron absorption by the moderator, then the corrected eqn. obtained by multiplying the solution by p is

$$q = \phi(r)\, \Sigma_a \varepsilon f \eta p\, e^{-B^2 r}$$

$$= k_\infty \Sigma_a \phi e(r)\, e^{-B^2 r} \quad \text{...(17)}$$

Substituting this value of q in eqn. (6), we get

$$\nabla^2 \phi - \frac{3}{\lambda_{tr}\lambda_a}\phi + \frac{3k_\infty e^{-B^2 r}}{\lambda_{tr}\lambda_a}\phi = 0$$

or using $\frac{1}{3}\lambda_{tr}\lambda_a = L^2$, we get

$$\nabla^2 \phi + \left(\frac{k_\infty e^{-Bh^2 r - 1}}{L}\right)\phi = 0 \quad \text{...(18)}$$

This is a general *diffusion eqn. for an actual reactor.*

Reactor Buckling

Using eqns. (7) and (17), we get

$$k_\infty \Sigma_a e^{-B^2 r} (\nabla^2 \phi + B^2 \phi) = 0,$$

or $$\nabla^2 \phi + B^2 \phi = 0 \qquad \text{...(19)}$$

Since the thermal neutron flux distribution ϕ (r) depends on the geometry of the assembly, the value of B^2 is similarly determined by the reactor geometry and is known as geometrical buckling, denoted by B_g^2. Combining eqns (18) and (19), we get

$$\left[\frac{k_\infty}{(1 + L^2 B^2)}\right] e^{-B^2 r} = 1 \qquad \text{...(20)}$$

This is the critical equation. The left hand side is also denoted by k_{eff}. The numerical value of B^2 depends upon the physical properties of the reactor material and is, therefore, called the *material buckling* (B_m^2) of the reactor. Geometrical buckling B_g^2 decreases as we increase geometrical dimensions of a critical reactor. But this increase results in a k_{eff} greater than unity. As material buckling B_m^2 does not change with the reactor size. Hence for a super critical reactor ($k_{eff} > 1$), B_m^2 must be greater than B_g^2 Similarly, a reduction in size makes the reactor subcritical ($k_{eff} < 1$) for which B_g^2 is greater than B_m^2.

Nonleakage Factors

Fraction of fast neutrons that does not leak out of the assembly during slowing down, l_f, is given by the ratio of the actual production rate of thermal neutrons over the maximum possible for a reactor of infinite size. Thus

$$l_f = \frac{k_\infty \Sigma_a \phi e^{-B^2 r}}{k_\infty \Sigma_a \phi} = e^{-B^2 r} \qquad \text{...(21)}$$

Thermal non-leakage factor l_{th} = 1—Thermal leakage factor

$$= 1 - \frac{\text{Thernal diffusion rate}}{\text{thermal diffusion rate} + \text{Thermal absorption rate}}$$

$$= 1 - \frac{-\frac{1}{3}\lambda_{tr}\nabla^2\phi}{-\frac{1}{3}\lambda_{tr}\nabla^2\phi + \Sigma_a\phi} = -\frac{\Sigma_a\phi}{-\frac{1}{3}\lambda_{tr}\nabla^2\phi + \Sigma_a\phi}$$

$$= \frac{1}{1 + \frac{1}{3}\lambda_{tr}\lambda_a B} = \frac{1}{1 + B^2 L^2} \qquad \text{...(22)}$$

$$\therefore \quad k_{eff} = k_\infty \, l_{th} l_f = k_\infty \, e^{-B^2\tau} / (1 + L^2B^2)$$

$$= \frac{k_\infty}{(1 + B^2\tau)(1 + L^2B^2)} = \frac{k_\infty}{1 + B^2(L^2 + \tau)} = \frac{k_\infty}{1 + B^2M^2} \quad ...(23)$$

Hence $L^2 + \tau$ is replaced by M^2, the migration area. M is known as migration length. Thus for large thermal critical reactor

$$k_{eff} = k_\infty/(1 + B^2M^2) = 1 \quad ...(24)$$

The shape of nuclear reactors are almost exclusively either a rectangular parallelepiped, a sphere, or a cylinder. The most suitable co-ordinate systems are cartesian for the first, spherical for the second and cylindrical for the last. We shall summarised here the solutions for these three co-ordinate systems.

Rectangular Parallelepiped of Sides a, b, c.

The equation of neutron flux in this reactor in cartesian co-ordinates is given by

$$\frac{\partial^2 \phi}{\partial x^2} + \frac{\partial^2 \phi}{\partial y^2} + \frac{\partial^2 \phi}{\partial z^2} + B^2\phi = 0 \quad ...(25)$$

The solution, the neutron flux, will be the function of x, y, z and can be written as the product of three functions as :

$$\phi = X(x)\, Y(y)\, Z(z). \quad ...(26)$$

$$\therefore \quad \frac{1}{X}\frac{\partial^2 X}{\partial x^2} + \frac{1}{Y}\frac{\partial^2 Y}{\partial y^2} + \frac{1}{Z}\frac{\partial^2 Z}{\partial z^2} + B^2 = 0 \quad ..(27)$$

Let us assume that

$$\frac{1}{X}\frac{\partial^2 X}{\partial x^2} = -\alpha^2, \; \frac{1}{Y}\frac{\partial^2 Y}{\partial y^2} = -\beta^2 \text{ and } \frac{1}{Z}\frac{\partial^2 Z}{\partial z^2} = \gamma^2, \quad ...(28)$$

$$\therefore \quad \alpha^2 + \beta^2 + \gamma^2 = B^2 \quad ...(30)$$

The solution of equation $\left(\frac{1}{X}\right)\frac{\partial^2 X}{\partial x^2} = -\alpha^2$ is supposed to be

$$X = A_1 \cos \alpha x + A_1' \sin \alpha x \quad ...(30)$$

Because sine function is not a symmetric function, hence the arbitrary constant A_1' has to vanish. Using now the boundry condition $\phi = 0$ for $x = \alpha/2$, we get $A_1 \cos a\, \alpha/2 = 0$ or $\alpha = \pi/a$.

Thus the solution becomes

$$X = A_1 \cos (\pi/a)\, x \qquad ...(31)$$

Similarly we have $$Y = A_2 \cos (\pi/b)\, y \qquad ...(32)$$

and $$Z = A_3 \cos (\pi/c)\, z \qquad ...(33)$$

$$\phi = A \cos (\pi/a)x \cos (\pi/b)y \cos (\pi/c)z \qquad ...(34)$$

and $$B^2 = (\pi/a)^2 + (\pi/b)^2 + (\pi/c)^2. \qquad ...(35)$$

It can be shown that the minimum critical volume of a parallelepiped is a cube (a = b = c) for which $B^2 = 3\pi^2/a^2$.

$$\therefore \quad V_{min} = a^3 = 3^{3/2} + \pi^3/B^2 = 161/B^3 \qquad ...(36)$$

Spherical Reactor of Radios R

Because of spherical symmetry, if we take the origin of the reference system at the centre of the sphere, the solution will be independent of θ and ϕ. The diffusion equation thus can be written as

$$\frac{\partial^2 \phi}{\partial r^2} + \frac{2}{r}\frac{\partial \phi}{\partial r} + B^2\phi = 0 \qquad ...(37)$$

The solution of this equation is

$$\phi = (A_1/r) \sin Br + (A_2/r) \cos Br. \qquad ...(38)$$

With the use of the boundary conditions : ϕ is finite at r = 0 and is zero at r = B, we get $A_2 = 0$ and $BR = \pi$ or $R = \pi/B$. Thus the solution for the spherical reactor is

$$\phi = (A_1/r) \sin (\pi/R)r. \qquad ...(39)$$

Under the critical condition the minimum volume

$$V_{min} = \frac{4}{2}\pi R^2 = \frac{4}{3}\pi\left(\frac{\pi}{B}\right)^2 = \frac{130}{B^2} \qquad ...(40)$$

Cylindrical Reactor

The diffusion equation in cylindrical co-ordinates with cylindrical symmetry (term containing derivative of angular co-ordinates $\theta = 0$) is

$$\frac{\partial^2 \phi}{\partial r^2} + \frac{1}{r}\frac{\partial^2 \phi}{\partial r} + \frac{\partial^2 \phi}{\partial z^2} + B^2\phi = 0 \qquad ...(41)$$

The solution of this equation can be written as the product of two functions as

$$\phi = R(r)\,Z(z) \qquad ...(42)$$

Thus equation (41) can be written as

$$\frac{1}{R}\frac{\partial^2 R}{\partial r^2}+\frac{1}{rR}\frac{\partial R}{\partial r}+\frac{1}{Z}\frac{\partial^2 Z}{\partial z^2}+B^2=0 \qquad ...(43)$$

Let
$$\left(\frac{1}{Z}\right)\frac{\partial^2 Z}{\partial z^2}=-\alpha^2 \qquad ...(44)$$

The solution is $Z = A_2 \cos \alpha z + A_2' \sin \alpha z$.

To have cylindrical symmetry A_2' has to vanish because since function is not a symmetric function. We then have

$$Z = A_2 \cos \alpha z \qquad ...(45)$$

By using boundary condition $\phi = 0$, for $z = H/2$, we have

$$A_1 \cos \alpha H/2 = 0$$

$$\therefore \quad \frac{\alpha H}{2}=\frac{\pi}{2} \quad \text{or} \quad \alpha=\frac{\pi}{H} \qquad ...(46)$$

Hence
$$Z = A_2 \cos (\pi/H/)z. \qquad ...(47)$$

Substituting the value of a and hence that of term containing Z in equation (43), we have

$$\frac{d^2R}{dr^2}+\frac{1}{r}\frac{dR}{dr}+\left(B^2-\frac{\pi^2}{H^2}\right)R=0 \qquad ...(48)$$

Writing $B^2 = \frac{\pi^2}{H^2} = K^2$ and multiplying by r^3, we have

$$r^2 = \frac{d^2R}{dr^2}+r\frac{dR}{dr}+K^2r^2R=0$$

Calling $rK = \mu$, we can write finally

$$\mu^2 = \frac{d^2R}{d\mu^2}+\mu\frac{dR}{d\mu}+\mu^2R=0 \qquad ...(49)$$

It is a linear differential equation of second order in cylindrical co-ordinates and known as the Bessel's eqn. of zero order. Its solution is the Bessel functions of zero order and can be written as

$$R = A_3 J_0(\mu) \qquad ...(50)$$

Combining the results we have finally

$$\phi = AJ_0\,(Kr)\cos(\pi/H)z. \qquad ...(51)$$

From the Bessel function tables of zero order we find that $J_0(Kr)$ reaches the value 0 when Kr = 2.405. It corresponds to r = radius R of the cylinder. Hence K = 2.405/R.

Substituting this value of K in equation (51), we have

$$\phi = AJ_0\left[\left(\frac{2.405}{R}\right)r\right]\cos\left(\frac{\pi}{H}\right)z \qquad ...(52)$$

$$\therefore \quad B^2 = \frac{\pi^2}{H^2 + (2.405/R)^2} \qquad ...(53)$$

For the smallest critical volume of the cylinder V = ($\pi R^2 H$) should be minimum. Hence

$$\frac{dV}{dH} = \frac{d}{dH}\left(\pi \times \frac{H^2(2.405)^2}{B^2H^2 - \pi^2} \times H\right) = 0$$

or $$B^2 = \frac{2\pi^2}{H^2} \qquad ...(54)$$

Therefore, $$V_{min} = \frac{\pi(2.405)^2}{2\pi^2}\left(\frac{\sqrt{3}\pi}{B}\right)^3 = \frac{148}{B^3} \qquad ...(55)$$

From the above results it is apparent that for a nuclear reactor using the same materials the most convenient shape is the spherical one as this shape requires the minimum volume. It is also clear from above result that the minimum volume of any of the shapes considered is proportional to the inverse of tie coefficient B^3.

The quantity B depends upon the average number of neutrons created per fission, the diffusion length and the slowing down length. As these depend upon reactor material, the quantity B is known as the *material buckling* and denoted by B_m.

Hence the minimum volume is defined as the critical condition under which geometrical buckling is equal to the material buckling. This determines the amount of material required for the reactor to become critical.

CLASSIFICATION OF REACTORS

Reactors may be classified in a wide variety of ways. For example they may be classified by the manner in which the fuel and the moderator are mixed. In a *homogeneous reactor* the fuel and moderator arc mixed in the form of a solution. *Most of the nuclear* reactors are of the heterogeneous type, in which fuel is concentrated in plates, rods, or hollow cylinders, which are distributed in a regular pattern (lattice) within the moderator. Reactors are further classified according to the energy of the neutrons that causc fission. These reactors ire known as *thermal, intermediate* or fast according to the thermal neutrons (0.025 eV), intermediate neutrons (1 to 10 keV) and fast neutrons (above than 10 keV) respectively. Reactors are also classified according to their purpose. The types to be examined here are (a) research reactors, (b) production reactors, and (c) power reactors.

(a) Research Reactors

Research reactors are used primarily to supply neutrons for physical research and radioisotope manufacture. In these reactors the total energy liberated is comparatively small, since they operate at a low level of reactivity. In these cases cooling is required only to prevent over heating. We shall describe here five main types of research reactors :

1. Graphite-Moderated. Research Reactor

The graphite-moderated, heterogeneous, natural uranium reactor has the distinction of being the first reactor built and operated. This first reactor, now called CP-1 (Chicago Pile-1), was put into operation in 1942. This was operated without coolant flow at a power level of 200 watts, Similar assemblies with air as coolant, like the GLEEP (graphite low energy experimental pile) and BEPO (British experimental pile) at Harwell (England) and the X—10 Clinton pile at Oak Ridge and BNL (Brookhaven National Laboratory) are well known examples as they have been the work horses of reactor technology and research. All these reactors use fuel in the shape of cylindrical rods, or variations on this shape, clad usually in aluminium. "Several hundred tons of graphite are used as moderator and reflector, with several hundred cylindrical channels passing right through the graphite to allow the introduction and positioning of the fuel elements. Cooling gas is made to pass over the elements through the channels at low pressure. The power and neutron flux for different reactors are given in Table 1.

Table 1 : Graphite—Moderated, Air-coolant Heterogeneous, Research Reactor

Name	Size	Fuel and its arrangement	Reflector	Power (kW) Neutron Flux
CP-1	9'×9'×12'	Uranium (6 tons) its oxide (40 tons) spherical lumbs $\left(D = 2\frac{1}{4}\right)$ at $8\frac{1}{4}$ cubical spacing	1 ft.	0.2
CP-2	18'×19'×20'	Uranium (10 tons) its oxide (40 tons) similar to CP-1	"	0.2–2 3×10^3
GLEEP	Cylinder D = 9.5' H = 17'	Uranium (33 tons) slugs D=0.97", H=12", $7\frac{1}{4}$ spacing, Al-clad	2ft.	100 3.7×10^{10}
BEPO	Cylinder D = 20 H=20'	Uranium (40 tons) slugs (D=0.9", H=12", $7\frac{1}{4}$ spacing. Al-clad	3ft.	4000-6000 10^{12}
X-10	20' Cube	Uranium (49 tons) slugs (Dù1.K//SS4" spacing Al-clad.	2ft.	1000-3800 $\sim10^{12}$
BNL	25' Cube	Uranium (100 tons) slue (D=1.1", H=4"), 8" spacing Al-clad	4,5'	30000 4×10^{12}

2. Water Boiler Type Reactor

Perhaps one of the roost misleading names of a reactor type is the water boiler. The water boiler is usually a homogeneous mixture of a highly enriched uranium salt dissolved in ordinary water. The name water boiler often leads to a mistaken belief that these reactors are the boiling water type of reactor. Actually the boiling water reactors arc quite different and are under the bead "power reactors". The name arises from the fact that a sudden increase in power will cause the formation of steam bubbles in the solution, which in turn will shut down the reactor quickly. Under normal conditions, the solution does not boil as its temperature is kept below 80°C by circulating water through the coils inside of the core vessel.

The first reactor of this type was built as Los Alamos in early 1944. It was named LOPO (low power) because of its low power operation. It was improved in December for high power operation and was named HYPO (high power). It was further improved for superpower, named SUPO (Super power), and is in operation since 1951. Another water boiler of note is the reactor built at the North Carolina State College in 1953. It is the first reactor owned and operated by a University anywhere in the world. Similar assemblies, like WBNS (water boiler neutron source) California and ARR (Armour research reactor) Chicago and KEWB (kinetic experimental water boiler) California are well known examples. Few properties of these reactors are given in Table 2.

Table 2 : Water Boiler, Water Coolant, Water Moderator, Homogeneous Research Reactor

Name	Size	Fuel	Reflector	Power (kW)/ Neutron flux
LOPO	Sphere 1'dia 15 litre	Enriched uranyl sulphate solution; 15 litres; U^{238} (580 gm.), U^{238} (3380 gm.).	BeO and Graphite	5×10^{-5}
HYPO	"	Enriched U-nitrate solution, U^{235} (897 gm.) U^{238} (5341 gm.).	BeO and Graphite	63×11^{11}
SUPO	"	90% enriched U-nitrate, 0.8 kg. in 13 litres.	Graphite	35-45 1.7×10^{12}
WBNS	"	90% enriched uranyl nitrate solution, U^{235} (700 gms.).	Graphite	0.005 2×10^{5}
North Carolina	Cylinder $(D=10\frac{3}{4}$, H=11')	90% enriched UO_2 SO_4 solution,. U^{235} (848 gm.) in 14 litres.	Graphite	10 5×10^{11}
ARR	15 litre spherical core	88% enriched uranyl sulphate.	Graphite	50 10^{12}
KEWB	Sphere (1,1/4, dia)	90% enriched uranyl sulphate, U^{235} (1750 gm.).	Graphite	50 9×10^{11}

The water boiler reactor has proved to be versatile research tool providing intense sources of neutrons and γ-rays. It has a great merit of simplicity of design and construction. Its use is somewhat limited because of its low power rating.

3. Swimming Pool Reactor

The turn swimming pool reactor is the popular name of light water moderated heterogeneous reactor which consists of a lattice of enriched uranium fuel immersed in a large pool of water. The water in the pool plays the role of moderator, coolant and shield. The chief advantages of this type of reactor are simplicity, low cost, safety, flexibility and accessibility. The first swimming pool reactor was completed in 1950 at the Oak Ridge National Laboratory. It was named BSR because was designed with the view of allowing radiation effects on bulky materials to be observed by immersing them in the water surrounding the core.

The Indian Reactor **Apsara** belongs to this type. This reactor was completed on August 4, 1956 and was the first reactor to go into operation in Asia. It was named Apsars on the name of Legendary sea mermaid because it is half under the tea and half outside. It was designed and built by the Indian engineers and scientists under the guidance of the late Dr. H.J. Bhaba. This reactor is designed for a maximum power level of 1 mw and contains large quantity of light water in a concrete pool. The fuel assembly is attached to steel frames and is suspended from above from an adjustable bridge placed on rails.

This bridge spanes the pool. The fuel elements are arranged in the rectangular assembly. These are made of enriched uranium alloy and arc held inside aluminium cans. Few control rods of Cd metal are also introduced in the fuel assembly and are adjusted automatically so that reactor does not become super critical. Holes are provided through graphite reflector through which thermal neutrons are available. When the reactor becomes critical the water surrounding the core emits a brilliant Cerenkov light glow, due to the passage of energetic electrons through water, can be seen from above the pool in the broad day light as well.

The first fuel charge was removed from the reactor in Dec. 1964 since it had attached its maximum burn up. Maintenance of its various components/improvements in cooling and control systems were carried out in the first quarter of 1965. It was started up with the second stage of fuel elements, obtained from U.K., in July 1965. The standard element of second fuel charge contains 140 gm of U^{235} (80% enriched uranium). For normal operation the reactor core contains, about 25 fuel elements- This reactor has been used for the production of isotopes, irradiation of biological samples, and for testing the various types of reactors and health physics instruments. It has been used for fundamental research in nuclear physics and structure of materials. A large number of isotopes such as P^{32}, S^{35}, I^{131} and Au^{198}, produced in this reactor, are used in research instruments and hospitals.

Table 3 : Swimming Pool Water Coolant, Water Moderator, Heterogeneous Research Reactor

Name	Size	Fuel	Reflector	Power (kW) Neutron flux
BSR	20'×20'×40'	MTR type 3.5 kg. U^{235}, Fuel core 1'×1'×2'	H_2O or BeO	100-1000 2×10^{13}
NRLR (Naval)	$49' \times 32' \times 28\frac{1}{2}$	MTR type, Fuel core 21"×21"×24'	H_2O or Graphite	100-1000 10^{12}
Baitelle RR		MTR type 2.8 kg. U^{235}	H_2O or Graphite	1000 1.4×10^{13}
Swiss Reactor	10' dia 2' deep	23 MTR type fuel assemblies.	H_20	10-100 10^{11}
Melusine (France)		Enriched uranium	H_2O	1000 10^{13}
Apsara (India)		Enriched uranium alloy	Graphite	100-1000 10^{12}

There are about two dozen swimming pool reactors available in the world. Few properties of some reactors are given in Table 3.

4. Light-Water-Moderator, Tank type Reactor

This reactor is very similar to pool reactor, in principle, except that the core is suspended in a deep cylindrical tank of water and cannot be moved and it preferred over 'one of the pool type when a large number of holes's for neutron beams in many different directions is an important requirement.

A reactor with increased neutrons flux was designed and built in 1952 at the National Reactor Testing Station in Arco, Idaho and was named Material Testing Reactor (MTR).

Table 4 : Light water moderator. Water Coolant, Heterogeneous, Tank Type Reactor.

Name	Size	Fuel	Reflector	Power (kW) Neutron flux
MTR	Overall (32'×34'×40') reactor tank (56' dia, 30' high), fuel assembly (16"×28"×24")	Sandwich type > 90% enriched fuel elements of U-Al alloy, 19 curved vertical fuel plates in 3×3×24 inches box 4.5 kg. U^{235} in Al tank, 54" dia with side extension to form water well	Be and graphite	4×10^{1} 3.7×10^{14}
ETR	32" dia 36" high	49 MTR type fuel elements	Be and Graphite	1.75×10^{5} 3×10^{11}
WTR	Core (D=28", H=36")	60 MTR type fuel assemblies	H_2O	10^{14}
HFR (Highflux)		90% enriched MTR type fuel elements U^{235} (4.2 kg.)	H_2O	2×10^{14} 4×10^{11}

The MTR fuel elements are heterogeneous plate type elements. These plates, often referred to as sandwich type plates, contain an inner core of uranium-aluminium alloy which Is 0.02" thick. This core is clad (bread) on either side by 0.02" thick aluminium. Eighteen of the plates are brazed together in aluminium side plates with a spacing of 0.117" between plates to form which is generally known as the MTR type element. Each one of the MTR elements contains 140 gm. of U^{235} and the critical mass for the reactor is approximately 3 kg. The properties of few other reactors of this type are listed in the Table 4.

5. *Heavy-Water-Moderator, Tank type Reactor*

The desirability of obtaining heavy water for use as a neutron moderator lies in the extremely small thermal absorption cross section deuterium. Because of this property the neutron economy in a heavy water system is greatly improved. The first heavy water moderator reactor was the Chicago Pile 3 (CP-3) which went into operation on may 1944 at the Argonne National Lab. and was dismantled in January 1950. It is a simple tank type of reactor. Canada has specilized in heavy water natural uranium reactor because of the abundant supply of natural uranium and no facilities for enrich ment of natural uranium. The first reactor constructed in Canada was completed in Sep. 1945.

Cirns : formerly known as CIR, is India's second research reactor, built as a joint Indo-Canadian project under the Colombo Plan at Trombay. This reactor was completed on 10 July, 1960. The name CIR stands for Canadian Indian Reactor and the new name cirus is the name of the star close to our solar system. It is a natural uranium heavy water moderated and light water cooled high flux research reactor with a terminal power of 40 MW. It is Operated round the clock. The standard fuel rod contains about 55 kg. of natural uranium. For normal operation the reactor core contains about 185 fuel rods and 23 tonnes of heavy water. This reactor is used for basic and applied research in physics, chemistry, biology, metallurgy and reactor technology, for the production of large quantities of high activity radio isotopes as well as plutonium-239 and uranium-233.

Zerlina : Is India's third research reactor. It attained criticality on Jan. 14, 1961. The name zerlina stands for *zero energy lattice investigation nuclear assembly* as it was made to study the lattice assemblies of different kinds. In this reactor natural uranium is used as a fuel and heavy water as a moderator. The standard fuel rod contains about 45 kg. of natural uranium. The standard core contains 80 fuel rods and about 8 tonnes of heavy water. It is operated on a two shift basis upto a maximum power of 100W and is being extensively used for physics and studies of uranium—heavy water lattice.

Purnima : It is India's first experimental zero energy fast reactor. It attained criticality on May 22, 1972. It uses plutonium oxide as fuel. It is used for the training of personnel in the operation of plutonium fuelled fast reactors. Experiments were carried out by scientists to measure parameters used in the construction of the core of the pulsed fast reactor at Kalpakham.

R-5 : It is a thermal research reactor with a nominal power of 100 MW and is located adjacent to 40 MW cirus reactor.

The natural uranium fuel elements of cirus and zerlina and the plutonium oxide for purnima are fabricated in Trombay. This fabrication facility went into operation in June 1959. The present capacity of this plant is about 30 to 44 tones of fuel elements per year.

(b) Production Reactors

The purpose of a production reactor to convert fertile into fissile material. The fissible materials are used as a fuel in other reactors. In

Hanford works production reactors, the natural uranium is used as a fuel, graphite as a moderator and water as a coolant. Because of the relatively large number of neutrons captured by a hydrogen nuclei present in water, the proportion available for capture by U^{238} is reduced considerably. The efficiency of conversion of U^{238} into Pu^{239} is thus appreciably less than unity. The ratio is always smaller than 0.8 in this type of reactor.

The efficiency of the conversion process can be increased by decreasing resonance escape probability. It is mainly by capture of neutrons in the resonance region. U^{238} is converted into Pu^{239}. Since maximum value of multiplication factor in uranium graphite reactor is 1.07, hence any decrease in the resonance escape probability can make the reactor impossible to attain criticality. Multiplication factor would be a large as 1.25 if heavy water is used as a moderator. A substantial decrease in the resonance escape probability is thus permissible without making reactor impossible to attain criticality. Thus we see that conversion efficiency of a production reactor using heavy water as a moderator is greater than the reactor using graphite as a moderator.

(c) Power Reactors

The primary purpose of a power reactor is the utilization of the fission energy produced in the reactor core and to convert it into useful power. Heat generated in fission process is used, either directly, or indirectly, to produce steam at high temperature and pressure or to heat a compressed gas. The steam or heated gas is used as a working substance in a turbine. The turbine can be connected to an electric generator or can be used as a source of mechanical power. Choice of the heat exchange material is governed by three important considerations.

(d) It must have a high thermal conductivity so as to carry away the heat efficiently and give it up in a heat exchanger.

(b) It must have a low neutron capture cross-section so as not to upset the reactor characteristic.

(c) It must not be decomposed by intense radiation.

Now we shall discuss some of the reactor types which appear to show the greatest promise for power production.

Pressurized Water Reactor

This is a heterogeneous reactor that uses slightly enriched uranium (1.4 to 2% of U^{235}) as fuel and light water as a moderator and coolant.

This type of reactor has received much attention in the United Slates. The fuel elements are in the form of rods or plates. The core is contained in a vessel under a pressure of 1000 to 2000 pounds per square inch. The pressurized water is circulated through the reactor core, from which it removes heat, and then through on external heat exchanger. Heat exchanger is used to abstract heat energy from the coolant in order to make it available for conversion into other forms of energy. High pressure is maintained in the core in order to raise the boiling point of the coolant to high enough temperature only because a consequence of the laws of thermodynamics is that heat can be more coefficiently convened into useful energy if it is released at a high temperature.

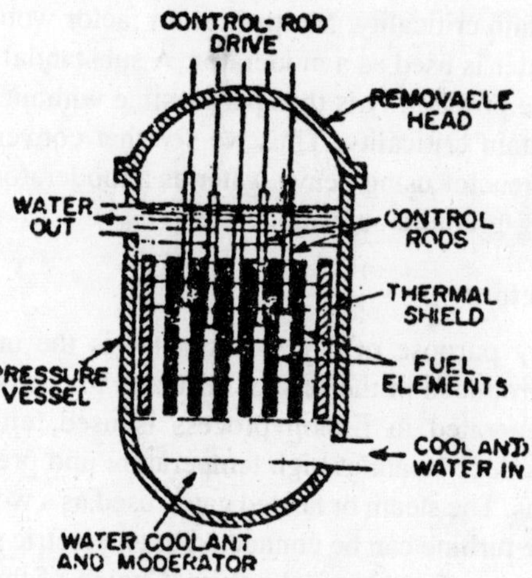

Fig. 1 : Schematic diagram of the pressure vessel and core of a pressurized water reactor

This first full scale nuclear power plant in the United States at Shipping port, Pennsylvania, was completed at the end of 1957. The first one produced originally 60 mega watts of electric power. This power was latter increased to 100 megawatts with a capability of going to 150 megawatts of electricity Several other reactors of this type have been (or are being) constructed in the United States and in other countries e.g., France, Italy and U.S.S.R.

Boiling water Reactor

The boiling water reactor is a small nuclear power plant designed to allow steam to be generated directly in the reactor core. This uses light water as moderator coolant. This reactor system is more advantageous than pressurized water reactor. The pressure in the reactor vessel does not have to be much greater than the required steam pressure while in PWR the pressures used to prevent boiling are significantly larger. External heat exchangers are also not required. Enriched uranium is used as a fuel.

The experimental boiling water reactor at Argonne National Laboratory was the first reactor of this type. It was designed in the end of the year 1956 and electricity at the power of 5 megawatts was produced. The first boiling water reactor to produce substantial amount of electric power was started up in 1960 at the nuclear power station, near Chicago, Illinois.

Central station power plants utilizing boiling water reactors are also in operation in other parts of U.S.A. as well as in Italy and Germany. The maximum power produced by most recent plants is 1000 megawatts of electricity and the steam produced is at a temperature of 285° and a pressure of about 1000 psi.

The 420 MW energy station, at Tarapur (India), consists of two enriched uranium reactors of the boiling water type. These reactors were built with the help of the Central Electric Company of the U.S.A. under a turn key contract, on the basis of specifications provided by the Tarapur Atomic Power Project of the Department of Atomic Energy. Nuclear power began to flow from the station when it became operational at 8.15 p.m. on April I, 1969. These reactors operated extremely well in the first few years. However, some problems were raised in the years thereafter. Now since 1975 the performance is rather good.

This type of reactors are particularly adapted to integral super-heating. The steam is produced in one part of the core and super-heated in another part. The first such plant, completed in 1964, was the Boiling Nuclear Superheat (BONUS) reactor in Puerto Rico. In this reactor water is boiled in the central region of the core and the steam is superheated in the outer region. The second superheat reactor, completed in 1965, is at the Pathfinder power station at Sioux Falls. Sout Dakota. The arrangement is reverse of that in the first one. The temperature of exist steam is 480° in the first case and 440°C in the second one.

Heavy Water Moderated Reactor

There arc two advantages of reactors moderated by heavy water as compared with ordinary water as a moderator. First, the total cost of the core and cost of the consumed fuel arc both less in the heavy water moderated reactor because natural uranium or very slightly enriched material is used as a fuel. Second, the conversion ratio for the regeneration of fissile material can be close to unity. Both factors decrease the cost of the power, produced. As the high price of the heavy water increases the cost of power, this type of reactor cannot be used for large plants with an electrical capacity of 500 MW.

The first proto-type of a heavy water moderated power reactor is the Nuclear Power Demonstration Reactor (NPDR) completed in 1962 at Canada. It is of the pressurized water reactor type with heavy water as moderator and coolant instead of the light water. The coolant leaves the reactor at 277° and produced steam at about 230° in the heat exchanger.

The first boiling water reactor which uses heavy water as moderator is the Halden (Norway) reactor. In this reactor heavy water coolant-moderator boils and produces heavy steam. It is only one of its kinds in existence and has a designed thermal powers of 20 megawatts.

The second nuclear power station (430 MW) near. Kola at Rana Pratap Sagar in Rajasthan employs heavy water moderated reactors using natural uranium as fuel. The first unit of the reactor became critical in 1972 and started feeding power in 1973 into the power system of Rajasthan. In this power station the design and bulk of the equipment have come from Canada, but the construction and installation activities were done by the Department of Atomic Energy. Unfortunately there have been failures in two stages due to deficiencies in design and manufacture of the turbine generator; designed in U. K. and supplied from Canada. For the second unit of this station under construction, a major step in indigenisation has been taken. The reactor vessel, shields, boilors and fuelling machines have been made in and India and the responsibility has been taken over by Indian Scientists and Engineers.

The third nuclear power station (470 MW) is under construction at Kalpakham about 100 km away from Madras. In this instance complete responsibility for design. Engineering and construction of the project rests with the Department of Atomic Energy. The design generally follows the design used at Rajsthan, significant changes have been made in increase the output of the units by about 10%. This is 80% Indian, only 20% components are from aboard. Apart from reactor vessel and shields, boilers and fuelling machines, the equipment manufactured in India include the turbine generators, zirconium alloy components and the entire instrumentation and control system. The total fuel charge as well as heavy water will come from Indian plants.

The fourth Indian Atomic power plant is under construction at Narora. It will have two atomic reactors of 235 MW each and well deliver 440 MW of power to the northern power grid when completed. It will use natural uranium at fuel and heavy water as moderator like the power stations at Ranapratap Bagh and Kalpakham. This power station will be 100% from Indian resources.

Organic Moderated Reactor

Another method for achieving high temperatures at moderate pressures in a thermal reactor is to moderase and cool with an organic liquid containing a high hydrogen content and of his high boiling point. The organic moderated Reactor Testing Experiment (OMRE), was completed in 1957 at the National reactor Testing Station. Arco, Idaho. The organic liquid used in this reactor is a mixture of di and ter-phenyle and has a normal boiling point at atmospheric pressure of about 340°C and 370°C at the operating pressure of some 200 psi. The temperature is limited to a maximum of 325°C because the thermal decomposition of the organic compounds become more than can be tolerated. The experience gained was utilized in the design of a small power plant, called the Piqua Nuclear Power Facility (PNPF) at Piqua Ohia. The coolant used is a mixture of the three terphenyls ($C_{18}H_{14}$) at a pressure of 120 psi. It leaves the reactor at a temperature of 302°1C. Superheated steam at 288°C and a pressure of 440 psi is produced in the heat exchanger. The electric output of this plant is 11.5 megawatts.

Gas-Cooled Reactor

Several of the early experimental reactors in the United States were cooled with forced-air. Until 1957, this type of reactors were not developed in the United States because of various reasons. In the mean time air was employed as the coolant in the graphite moderated production reactors at Windscale in U. K. The graphite moderated power reactor with nitrogen carbon dioxide as a coolant and natural uranium as the fuel have been constructed in England, France and other countries. The carbon dioxide gas at a pressure of about 150 to nearly 400 psi in the later designs is pumped through the reactor leaving at a temperature around 400°C. The superheated steam is generated, at temperatures in the approximate range of 320° to 390° in the heat exchanger About 300 megawatts of electricity is produced by most of the larger plants. Helium gas also used in some of the gas cooled reactors. In High Temperature Gas Cooled Reactor, completed in 1966 at Peach Bottom, Pennsylvania, the helium gas leaves the core temperature of 750°C and produces superheated steam at 538°C at and 1450 psi pressure. The use of beryllium oxide as a moderator in gas cooled power reactor was proposed by F. Daniels and his associates in 1944. With improvements in fabrication Experimental Beryllium Oxide Reactor was constructed but the work was terminated in 1966 because cost of fabrication is a significant disadvantage in power reactor with this moderator.

Sodium Graphite Reactor

The high boiling point of liquid sodium makes it an excellent cooled for a high temperature reactor. High temperatures can be achieved without high pressures. Sodium Reactor Experiment was developed to its critical state in 1957. In this reactor graphite was used as moderator, sodium as coolant and either natural or slightly enriched uranium as the fuel.

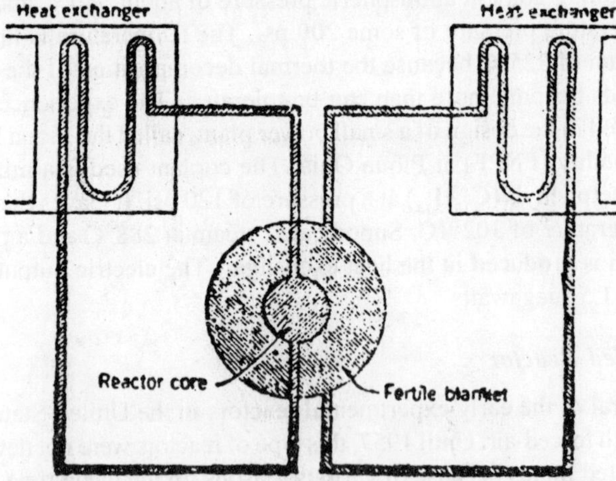

Fig. 2 : Schematic Diagram of homogeneous reactor.

Another reactor based upon SRE was completed in 1962. This is known as Hallam Nuclear Power Facility (HNPF). The temperature of the sodium leaving the reactor was 507°C and the steam was generated at a temperature of 446°C and a pressure of 850 psi. Its design power was 76 megawatts of electricity. It is very advantageous because sodium has excellent heat transfer characteristics and does not require excessive pumping power. After operating at full power for nearly two years, it was found that several of the stainless steel moderator cans had cracked and allowed sodium to leak. The cracking was caused by rupture of the steel due to the extended operation at high temperatures. In August 1965, the decision was made ultimately to close down the reactor.

Liquid Fuel Reactor

Aqueous homogeneous reactors are characterized by having the fissionable material such as uranyl sulphate or nitrate dissolved in light or heavy water and circulating this solution through a spherical vessel of stainless steel, in which chain reaction takes place and through a heat exchanger, in which the heat is transferred from hot solution to a working fluid and back to the vessel. The Homogeneous Reactor Experiment (HRE) at Oak Ridge National Laboratory was built in 1952 to test the feasibility of maintaining a fission chain at high temperatures and pressures. The advantages of an aqueous homogeneous reactor are the high power density, low fuel inventory, continuous removal of fission products and radiation damage products, high degree of nuclear stability, simple mechanical design and absence of structural materials in reactor core. Beside these advantages there are some drawbacks also.

(i) Great care must be taken to prevent leakage of highly radioactive solution at high temperature and pressure.

(ii) High pressures are required to produce high temperatures with the formation of steam in the reactor.

(iii) Radiolytic decomposition of H_2O or D_2O causes a number of problems.

Fast Reactor

The development of power reactors in which the bulk of fissions arc produced by high energy neutrons is of importance in the civilian power reactor development programmes. Such reactors not only produce useful power but also regenerate more fissile material than is consumed. Breeder reaction is prominent that is why it is also known as breeder reactor. The first Experi-mental Breeder was completed in August 1951 and produced over 100 watts of electric power on 20th December. It produced the first electricity from nuclear energy. In addition to the use of plutonium oxide and carbide, other alloy of plutonium, cerium and cobalt were also used as a fuel in designing of reactors.

CONTROLLED THERMONUCLEAR REACTIONS

In stars, nuclear fusion reactions release great amounts of energy. The rate depends upon the density and temperatùre of the gas and upon the cross sections or lifetimes of the reactions involved. Bearing in mind the masses of materials available on earth, it is certain that the reactions of the carbon cycle and the p-p chain would occur extremely slowly. There are thermonuclear reactions which occur much more rapidly and depend on an most abundant material. Among the nuclei of the hydrogen isotopes, ${}_1H^2$ and ${}_1H^3$ reaction are:

$${}_1H^2 + {}_1H^2 \rightarrow {}_1H^3 + {}_1H^1 + 4.0 \text{ Mev} \qquad \text{D-D reaction}$$

$${}_1H^2 + {}_1H^2 \rightarrow {}_2He^3 + {}_0n^1 + 3.2 \text{ Mev} \qquad \text{D-T reaction}$$

$${}_1H^2 + {}_1H^3 \rightarrow {}_2He^4 + {}_0n^1 + 17.6 \text{ Mev}$$

The D-D reaction can go in two equally probable ways, the first of which produces ${}_1H^3$ while the second produces ${}_2He^3$. When reactions take place in a chamber the deuterium can react with ${}_1H^3$ to give an α-particle and a neutron. 14.1 MeV energies are carried off by the neutron and 3.5 MeV by the α-particles. In these fusion reactions the energy released is much less than that released in fission reaction but the energy yield per unit mass of material is greater slightly. Deuterium occurs in nature with an abundance of about one part in 6500 hydrogen and can be separated from the lighter isotope quite cheaply It has been calculated that the energy equivalent of the deuterium in one gallon of water is the same as that obtained from the combustion of 300 gallons of gasoline. The more than 10^{20} gallons of water present in the oceans could thus supply the world's power requirement for several million years, if all the deuterium could be utilized to provide energy by fusion processes.

Thermonuclear (hydrogen) Bomb

Even before the achievement of atomic bomb, it was realized by J. Robert Oppenheimer and others that the explosion of an atomic bomb might create the high temperature needed to start a thermonuclear reaction and thus an atomic bomb might serve as the *fuse or trigger* to set off a hydrogen bomb. Hydrogen bomb can be made with deuteron or tritium or the combination. It is possible to produce tritium (as it occurs in nature only in infinitesimal amounts) by the neutron bombardment of lithium in a nuclear reactor. Calculations show that a given weight of tritium would release 7 times as much energy as an equal weight of Pu. One kg of tritium would release as much energy as 140,000 tons of T. N. T.

(the recognized abbreviation for the common chemical explosive 2, 4, 6-trinitrotoluene, assuming that the explosive energy of 1 ton of T. N. T. is 10^9 calories).

The energy produced in a hydrogen bomb is in an uncontrolled manner. To control thermonuclear reactions some of the essential conditions must be satisfied. The process must be self sustaining and the balance must be obtained between the energy released in fusion and the amount lost by radiation. The temperature corresponding to this condition of balance is known as critical ignition temperature.

The relation E = kT implies that it is of the order of 10^8 °K. At very high temperatures the atoms are fully ionized and these ions and the free electrons are moving about very rapidly. The result is a completely ionized gas, called a *plasma*.

The plasma is electrically neutral so that in the absence of electric or magnetic fields, there are no external forces acting on it except gravity. Because of the internal pressure, the plasma would expand in a vacuum to fill the container in which it is kept. When it comes in contact with the walls, cheats up the walls.

The power produced in fusion reactor will depend on the plasma density and temperature. Because of the very low density of the plasma, used in controlled thermonuclear reactions, and the relatively small volume of the system, it does not behave as a black body.

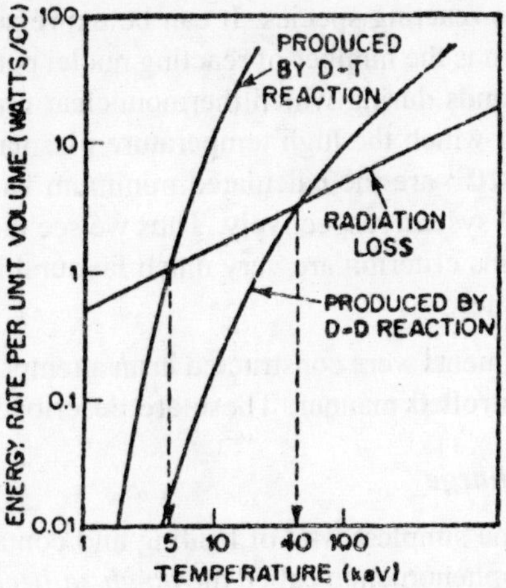

Fig. 3 : Critical ignition temperalure for the D-T and D-D reactions.

The radiation should consist mainly of bremsstrahlung accompanying the deflection of the rapidly moving electrons in the plasma by the electrostatic fields of the positively charged nuclei. The amounts of energies released per unit time per unit volume in the D-D reactions and in the D-T reactions and of energy lost as radiation at various temperatures.

In common practice the temperatures are expressed in keV, which is equivalent to 1.16×10^{7} °K It is clear from the figure that the bremsstrahlung loss equals to power produced at plasma temperature about 5 keV for a D-T and about 40 keV for D-D reactions This does not necessarily mean that the plasma temperature for a D-D reactor has to be higher than 40 keV to produce a net power gain.

The reason is that the bremsstrahlung is not lost but absorbed by the walls of the plasma container and converted therefore into heat in the way similar to the K- E. of the reaction products. The minimum operating temperature for a given reactor will therefore depend on the efficiency with which heat can be converted into other forms of energy. A part of this energy should be taken back by the plasma again to keep it at the operating voltage.

J.D. Lawson in England in 1957 pointed out another necessary condition, known as the *Lawson Criterion*, for a self-sustaining thermonuclear system. According to this criterion for a fusion reactor the total useful recoverable energy shall be at least sufficient to maintain the temperature of the reacting species. It can be expressed in terms of the product nt, where n is the number of reacting nuclei per unit volume and/ is the time in seconds during which thermonuclear reaction takes place or the time during which the high temperature plasma can be confined. 6×10^{19} and 2×10^{21} are the calculated minimum values for nt for D-T system and D-D system respectively. Thus we see that critical ignition temperature and this criterion are very much favourable for D-T reactor than for D-D reactor.

Several equipments were constructed in an attempt to produce fusion reactions in a controlled manner. These are described as given below.

(a) Pinched Discharge

In principle the simplest way of heating and containing the plasma is to make use of a phenomenon called the *pinch, utilized* in U.S.A., U.K., and in the U.S.S.R. simultaneously and independently in the early 1950. In it a very strong electric current amounting to a million amperes is passed through a deuterium like gas at low pressure. This current heats the gas and ionizes it, thus converting it into a *plasma*. At the same time the current flow produces a magnetic field with its lines of force encircling the plasma. The pressure of this field then pinches the plasma into a narrow region in the centre of the containing vessel which may be a torus or a straight tube. In this manner the plasma is not only kept away from the walls but heated also by the current itself and by the compression produced by the magnetic field, the current generates.

Several experimental machines have been built, some on quite a large scale. Two machines name Zeta and Sceptre were designed by Harwell group and by associated electrical industries respectively. An entirely different approach to the confinement and heating of plasmas is by the method of *fast magnitude compression* also called the *theta pinch*, originated in 1957 at the Los Alamos scientific laboratory.

The apparatus consists of an open ended tube containing deuterium gas at low pressure surrounded by a single turn coil. An oscillating current of several million amperes is passed through the coil for a very short time by the discharge of a large bank of capacitors. The first half cycle of the oscillating magnetic field produces an ionized sheath of plasma which is rapidly compressed and partially heated.

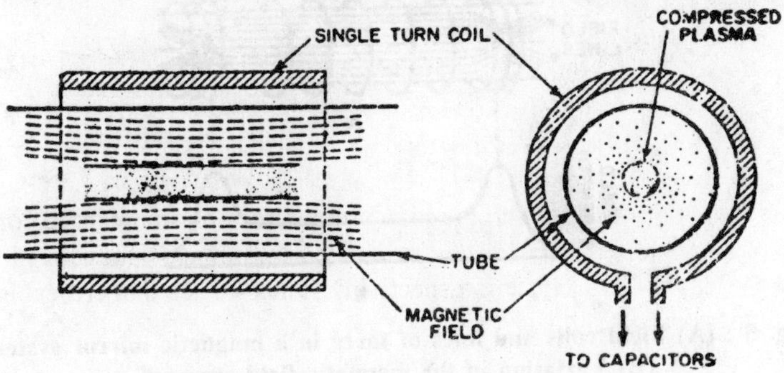

Fig. 4 : Schematic diagram of theta pinch apparatus.

In the second half cycle the direction of the magnetic field is reversed and the mutual annihilation of the oppositely directed fields releases a large amount of energy that heats the deuterium plasma. Scientists at Los Alamos were able to obtain plasmas with temperatures in excess of 5 keV and densities of about 4×10^{22} deuterons per m^3 by fast magnetic compression of low density preionized deuterium.

(b) Magnetic Mirror Systems

In the stellarator and other systems using closed containing tubes, the plasma can escape in the axial direction and so confinement is required only in radial direction. In magnetic mirror systems the plasma is confined in a *bottle shaped magnetic field.* This system consists of a straight tube with magnetic coils wound around it in such a way as to provide a field that is considerably stronger at the ends than in the middle. The ions move in spiral orbits with radii of curvature inversely proportional to the field strength and under certain conditions are reflected when move into regions of higher field strengths. In this way the magnetic mirrors inhibit but do not prevent entirely the escape of plasma from the ends of the tube. For a particle to be reflected it must have a significant component of velocity perpendicular to the magnetic field lines in the central region between the mirrors. The greater the value of B_{max}/B_{min}, the smaller the value of v_1 relative to actual "velocity v for which reflection is possible. The heating of the plasma in this system is accomplished by increasing the magnetic field relatively slowly such that the plasma is compressed. This compression can be carried out in several stages and the compressed plasma is transferred after each stage from one magnetic bottle to a smaller one in which it is further compressed.

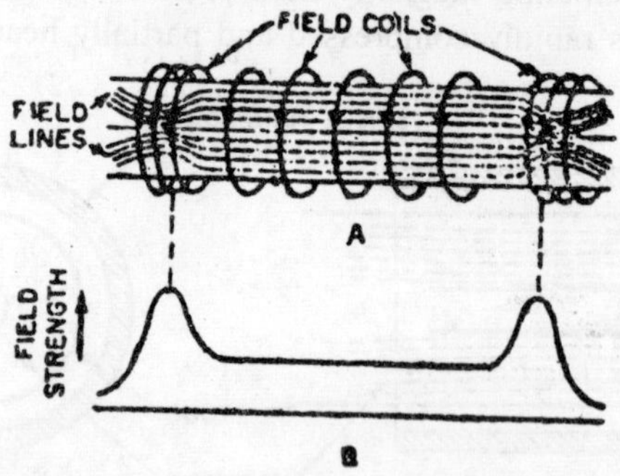

Fig. 5 : (A) Field coils and lines of force in a magnetic mirror system; (B) Variation in the magnetic field strength.

(c) Beam-injection Method

The injection of charged particles with a significant velocity component along the radial direction, directly across the magnetic field presents a difficulty. To solve this problem, the use of electrically neutral atoms of high energy was suggested, but it was very difficult to accelerate neutron atoms to a very high energy. Indirect process is the use of deuterium ions (~20 kV) which are allowed to pass into a chamber containing neutral deuterium gas. In this chamber a charge exchange occurs and the stream of high energy neutral deuterium atoms are produced, which then enter the magnetic field where they are ionized. Similar injection process used at the Oak Ridge National Laboratory starts with a high energy beam of molecular deuterium ions (D_2^+). These ions dissociate as $D_2^+ \rightarrow D^+ \rightarrow D^0$. Since the mass of deuterium is half that of the molecular ions, hence the resulting D^+ ions can be trapped but the undissociated D_2^+ ions are not. The neutral atoms are not affected by the magnetic field and hence escape. As a result of collisions, the trapped deuterons acquire random motion equivalent to a high temperature.The Effect of a Reflector. The critical size of a reactor core can be reduced by surrounding it with good moderating material. The latter has the properties of being a good neutron scatterer and does not usually absorb neutrons to any great extent. This material acts as a reflector. Graphite is often used for this purpose. The reactor with a reflector has certain distinct advantantages over the bare reactor which are stated as follows :

(a) Improved Neutron Economy

The reflector knocks neutrons which leave the reactor back into it and thus reduces neutron leakage from the core. The reflector also acts as a moderator for the fast neutrons that have entered it from the core. Since the absence of neutron absorbing material in the reflector reduces neutron loss due to resonance absorption, the moderation of fast neutrons in the reflector will be more efficient that in the core itself.

(b) Possibility of Fuel Saving

For constant core size reacton the improvement in the neutron economy reduces the amount of fuel or the fuel concentration required to achieve criticality. It means that reflector increases fuel saving. We also know that the reflector decreases the leakage or neutrons from the reactor and thus decreases the critical volume. Hence it is possible to reduce the size of the nuclear reactor including the reflector to below the size of a similar barer eactor with the sums fuel concentration.

(c) Improved Reactor Power Utilization

Since the power production rate is proportional to the average neutron flux, the reactor can be operated at a higher total power output for the same maximum neutron flux. The improvement in the power utilization is a consequence of the flattening of flux across the reactor core occuring due to the presence of the reflector. By virtue of this (flux flattening) effect the power production is also move uniform over the core volume.

NUCLEAR FISSION

When uranium was bombarded with neutrons it was found that a β-active product resulted, followed by a whole chain of β-active products such as

$$_{92}U^{38} + {_0n^1} \rightarrow {_{92}U^{239}} \rightarrow {_{93}X^{239}} \rightarrow {_{94}Y^{239}} \rightarrow \text{etc.}$$

It was recognised that more than one neutron was produced. Hahn and Strassman in 1939 proved that the product might be consisted of two large fragments, which they identified as barium and krypton. Frisch and Meitner in 1939 used the *word fission to describe the process which takes place when a heavy nucleus is caused to break down into two roughly equal parts, known as fission fragments.* In this process neutrons are also emitted with the release of considerable energy. This process has been also observed to occur when heavy nuclides are bombarded with protons, deuterons, α-particles and even electrons and gamma rays. Further work showed that lighter elements could also be fissioned by high energy particles, as for example in the case of copper

$$_{29}Cu^{63} + {_1p^1} \rightarrow {_{11}Na^{24}} \rightarrow {_{19}K^{39}} \rightarrow {_0n^1} \quad \text{...(1)}$$

Fission Types

(a) **Charged-particle Fission :** Elements with $Z > 90$ show fission process with protons, deuterons and α-particles. High energy charged particles induce fission in elements even in the middle of the periodic table.

(b) **Thermal Fission :** Since a 'thermal neutron adds negligible energy to the fissionable nucleus, it is clear from semi-empirical mass formula that the fission of the nuclei in which the compound nuclei are of even-even structure take place even with the thermal neutrons. Fission of U^{235} and Pu^{231} by thermal neutrons are the most important reactions.

(c) **Fast Fission :** Other isotopes of Uranium and other elements which form compound nuclei of even-odd structure enter into (n, f) reactions with fast neutrons (>I MeV). The example is U^{238}.

(d) Photo Fission : High energy photons induce fission in the heavier elements. 5.1 MeV y-rays can produce fission with U^{238}.

In all cases very large disintegration energies were released and fast neutrons are emitted. Meitner and Frisch indicated that, because of their exceptionally high neutrons to protons ratio, the fission fragments should be unstable, undergoing a chain of β^-distintegrations. For example

$$_{40}Zr^{98} \rightarrow {}_{41}Cb^{98} \rightarrow {}_{42}Mo^{98}$$

$$_{92}U^{235} + {}_0n^1 \rightarrow {}_{92}U^{236} \qquad\qquad + {}_{20}n^1 \qquad ...(2)$$

$$_{52}Te^{136} \rightarrow {}_{53}I^{136} \rightarrow {}_{54}Xe^{136}$$

A typical beta decay chain is

$$_{54}Xe^{140} \rightarrow {}_{55}Cs^{140} \rightarrow {}_{56}Ba^{140} \; {}_{57}La^{140} \rightarrow {}_{58}Ce^{140} \text{ (Stable)} \qquad ...(3)$$

For the nuclei near the middle of the periodic table (E_B = 8.5 MeV/nucleon), the fission process is expected to release a large amount of energy. Assuming fission of a ucleus of mass number 240, for which $E_B \simeq 7.6$, into two similar fragments with A ~ 120, the energy released in fission is 2 × 120 × 8.5 – 240 × 7.6 = 216 MeV. Assuming their separation equal to the sum of their nuclear radii, the Coulomb repulsion energy is given by

$$E_c = \frac{Z_1 Z_e e^2}{4\pi \epsilon_0 d} = \frac{52 \times 40 \times (1.6 \times 10^{-19})^2 \times 9 \times 10^9}{1.5 \times 10^{-14} \times 1.6 \times 10^{-13}} \text{ MeV.}$$

$$\sim 200 \text{ MeV.}$$

Thus we see that the Coulomb repulsion energy is nearly equal to the energy released in fission process. We can, therefore, regard this fission process as a result of Coulomb repulsion. The Q value of fission reaction (2) is also calculated from the exact mass difference of the two sides of the equation. The combined isotopic masses before and after fission are $\Sigma m_i = m(U^{235}) + m({}_0n^1) = 235.0439 + 1.0087 = 236.0526$ mu and $\Sigma m_f = m(Mo^{98}) + m(Xe^{136}) + m(2{}_0n^1) = 97.9054 + 135.9072 + 2.0154$ = 235.83 mu. Hence $Q = \Delta mc^2 = (\Sigma m_i - \Sigma m_f)c^2 = 210$ MeV, which is comparable with the value calculated by the binding energy method.

Deformation of Liquid Drop

The fission process can be explained with the help of liquid drop model. The incident neutron combines with the nucleus to form highly energetic compound nucleus. Its extra energy is partly the kinetic energy of the neutron but largely the added binding energy of the incident neutron. This energy appears to initiate a series of rapid oscillations in the drop, which tend to distort the spherical shape so that the drop may become ellipsoidal in shape. *The surface tension forces lend to make the drop return to its original spherical shape, while the excitation energy fends to distort the shape still further.* If the excitation energy is sufficiently large, the drop may attain the shape of a *dumb-bell.*

If the oscillations become so violent that stage fourth is reached then the final fission into stage fifth is inevitable. Thus there is a *threshold energy* or a *critical energy* required to produce stage fourth after which the nucleus can not return to stage first. When the distortion produced is not pronounced enough to get the nucleus beyond the critical point, the ellipsoid will return to the spherical shape with the excitation energy being liberated in the form of γ-rays and we have a radiative copture rather than fission. The potential energy of the drop in the different stages can be calculated as a function of the degree of deformation of the drop. It is plotted against r, the separation of the centres of two fission fragments. The curve is supposed to be divided into three regions.

In region I the fragments are completely separated and their potential energy E is simply the electrostatic Coulomb energy resulting from the mutual repulsion of the two positively charged nuclear fragments.

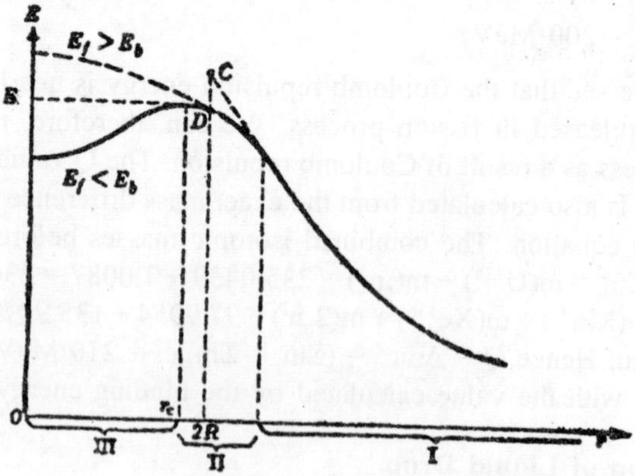

Fig. 6 : Potential energy curve for fission.

If distance r = 2R, when the drops just touch each other, energy E at that point if less than the corresponding Coulomb potential by an amount CD. This amount is equal to the potential of the surface forces which are just beginning to come into play at this point. As we pass through region II we reach the critical distance r_c, where the potential energy curve has a maximum value En.

This corresponds to barrier height and explains why fission does not take place spontaneously in all cases where $E_f > 0$. An additional amount of energy $E_a = R_b - E_f$, the activation energy is required by the nuclear system before the potential barrier can be surmounted and fission can take place. In the III region, the fragments have coalesced and the short range nuclear forces have become predominant.

Distribution of Fission Products

We have seen that a fissionable nucleus gives only two fission fragments which decay by β^-emission to a *stable end product*. Although, the sum of two large fragments always adds up to 234, there is a wide distribution in possible products.

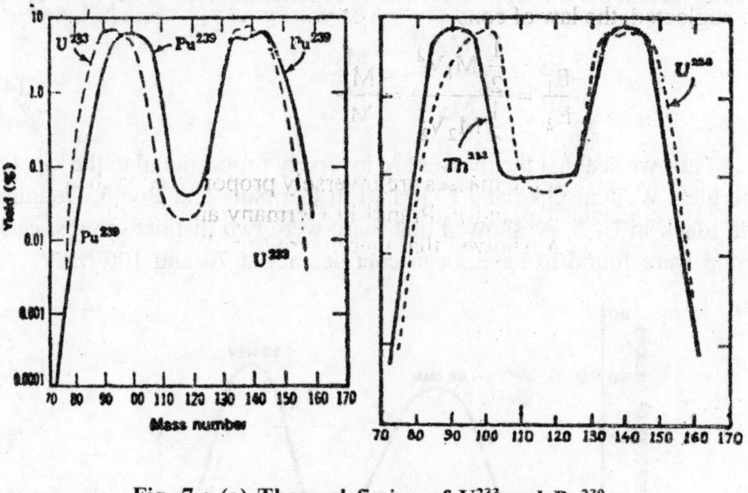

Fig. 7 : (a) Thermal fission of U^{233} and Pu^{239},
(b) Fast fission of Th^{232} and U^{238}.

The mass distribution of the fission products is shown most conveniently in the form of a *fission yield curve*, in which the percentage yields (in log scale) of the different products are plotted against mass number.

There is a tendency for masses to concentrate respectively round 90 and round 140. Actually about 97% of the total fission products fall within the narrow range 85—104 for the lighter fragments and 130-149 for the heavier. In the case of U^{235} the maxima lie near mass numbers 95 and 140; fission produced by slow neutrons is highly asymmetric process and division into two equal fragments occurs in only about 0.01% of the fissions.

Symmetric fission becomes increasingly more probable with increasing neutron energy. Single peak appears for high energy neutrons. The trough in the curve also tends to fill, when the nuclei are fissioned by the panicles other than neutrons, A comparison of the U^{233} and Pu^{239} fission yield curves shows that the distribution of the heavier products are practically the same, whereas the portions of the curves for the lighter fragment yields are displaced by six mass units with respect to other.

The mass distribution of the fission fragments can also be obtained from the distribution of their kinetic energies. The nucleus undergoing fission can be considered to be at rest initially and, if the neutrons emitted are neglected, the law of conservation of momentum gives $M_1V_1 = M_2V_2$.

$$\therefore \quad \frac{E_1}{E_2} = \frac{\frac{1}{2}M_1V_1^2}{\frac{1}{2}M_2V_2^2} = \frac{M_2}{M_1} \qquad ...(4)$$

Thus we see that the masses are inversely proportional to the kinetic energies. W. Jentschke and F. Prankl in Germany and Booth, Dunning and Slack in U. S. A. showed that there were two distinct groups, each group were found to have mean energies about 70 and 100 MeV.

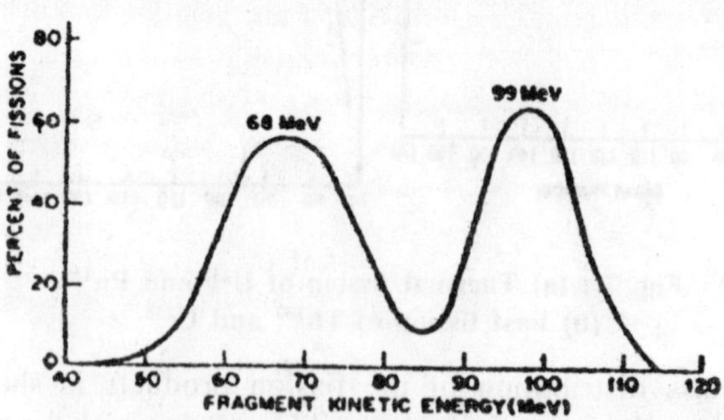

Fig. 8 : Kinetic energy distribution of fission fragments in U^{235}.

Later studies, based on measurement of ionization and velocities of the fission fragments, indicate that the K.E. of these fragments is 167 MeV in the fission of U^{235} by slow neutrons. There are two groups of energies, with maxima about 68 and 99 MeV. The difference between this quantity and the quantity released in fission process (200 MeV) is carried out by the gamma rays (11 MeV), neutrons (5 MeV), beta particles (7 MeV) and neutrinos (11 MeV).

From equation (4), it is clear that $\frac{M_1}{M_2} = \frac{E_2}{E_1} = \frac{99}{68} \simeq 1.5$, which is in close agreement with the ratio 140/95, Such measurement on the fission fragment energies gives the asymmetry of the fission process.

Neutron Emission in Fission

An accurate knowledge of the average number of neutrons emitted per second is of great importance to the scientists. Actually the number of neutrons released in any one fission is an integer, but average value, v_{av} of the number of neutrons in one fission is not an integer, because the fissionable nucleus can divide in at least 30 different ways, and can be obtained by the relation

$$v_{av} = \Sigma v_0 \, n \, v_0 / \Sigma n v_0 \qquad \text{...(5)}$$

Most of these neutrons are emitted within possibly 10^{-13} sec. and are called prompt neutrons. A smaller number of neutrons is also emitted with a time lag of several seconds to more than a minute after the fission. These neutrons arc called *delayed neutrons*. For sufficiently large piece of fissionable substance, the neutrons that are released in a first fission process will be absorbed by the other nuclei and produce new processes which in turn emit new neutrons. Actually some of the compound nuclei decay to the ground state by gamma emission, rather than fission. The ratio of the radiative capture cross-section to the fusion cross-section is usually denoted by a given by

$$\alpha = \frac{\sigma_r}{\sigma_f} \qquad \text{...(6)}$$

and the number of fusion neutrons released per neutron absorbed in the fissionable nuclide is denoted by η and is given by

$$\eta = \frac{v_{ab}}{(1+\alpha)} \qquad \text{...(7)}$$

The relative probability that a compound nucleus decays by fission is thus be $(1+\alpha)^{-1}$. Values of v_{av}, α and η for thermal neutrons are listed in the Table 5.

Table 5

	U^{233}	U^{235}	Natural uranium	Pu^{239}
σ_f	525 ± 4	577 ± 5	4.18 ± 0.06	742 ± 4
σ_r	53 ± 2	101 ± 5	3.50	286 ± 4
v_{av}	2.51 ± 0.02	2.44 ± 0.02	2.47	2.89 ± 0.03
α	0.101 ± 0.004	0.18 ± 0.01	0.83	0.39 ± 0.03
η	2.28 ± 0.02	2.07 ± 0.01	1.34 ± 0.02	2.08 ± 0.02

Fissile and Fertile Materials

For a number of reasons, isotopes such as U^{233}, which only fission with energetic neutrons, can not alone be used to fuel nuclear reactors. The only neutrally occurring nuclide that can be fissioned with thermal neutrons is U^{235}, which is 0.71% of the naturally occuring uranium. The only other nuclides that can undergo fission with thermal neutrons are U^{233} and Pu^{239}. These do not occur in nature but can be produced by the interaction of neutrons with Th^{232} and U^{238} respectively and are called *fissile materials*. Th^{232} and U^{238} are not fissile materials but can be used as raw material for the production of fissile isotopes and are called *fertile or fissionable materials*. The nuclear reactions which convert these fertile materials into fissile materials are called *breeding reactions*. They are neutrons capture processes with subsequent β-decay.

$$
\begin{aligned}
{}_{92}U^{238} + {}_{0}n^{1} &\rightarrow {}_{92}U^{230} + \gamma \\
{}_{92}U^{239} &\rightarrow {}_{93}Np^{239} + {}_{-1}e^{0} \\
{}_{93}Np^{239} &\rightarrow {}_{94}Pu^{239} + {}_{-1}e^{0}
\end{aligned} \qquad \text{...(8)}
$$

Similarly

$$
\begin{aligned}
{}_{90}Th^{232} + {}_{0}n^{1} &\rightarrow {}_{90}Th^{233} + \gamma \\
{}_{90}Th^{239} &\rightarrow {}_{91}Pa^{233} + {}_{-1}e^{0} \\
{}_{91}Pa^{233} &\rightarrow {}_{92}U^{233} + {}_{-1}e^{0}
\end{aligned} \qquad \text{...(9)}
$$

Spontaneous Fission

Most heavy nuclides undergo spontaneous fission in competition with the a-emission. Spontaneous fission is predicted by the empirical nuclear mass equation. Consider the special case, when the nucleus splits into equal parts. Neglecting the pairing term δ, we have the Q-value for the fission reaction

$$E_f = [{}_ZM^A = 2\,({}_{Z/2}M^{A/2})]c^2 \qquad ...(10)$$

Consider Weizsaker's semi empirical binding energy equation

$$M(Z, A) = ZM_P + A - Z)\,M_n - a_vA + a_sA^{2/3} + a_cZ^2A^{1/3} + a_a(A-2Z)^2A^{-1} \qquad ...(11)$$

$$M\left(\frac{1}{2}Z, \frac{1}{2}A\right) = \frac{1}{2}ZM_P = \frac{1}{2}(A-Z)M_n - a_c\left(\frac{1}{2}A\right) + a_s\left(\frac{1}{2}A\right)^{2/3} + a_c\left(\frac{1}{2}Z\right)^2\left(\frac{1}{2}A\right)^{-1/3} + \frac{1}{2}a_c(A-2Z)^2A^{-1} \qquad ...(12)$$

Substituting value of ${}_ZM^A$ and ${}_{Z/2}M^{A/2}$ in eqn. (10), we have

$$E_f = \left[a_s\left\{A^{2/3} - 2\left(\frac{1}{2}A\right)^{2/3}\right\} + a_c\left\{Z^2/A^{1/3} - 2\left(\frac{1}{2}Z\right)^2 / \left(\frac{1}{2}A\right)^{1/3}\right\}\right]c^2$$

$$= -3.42\,A^{2/3} + 0.22\,\frac{Z^2}{A^{1/3}}\ \text{MeV} \qquad ...(13)$$

This eqn shows that the splitting of a nucleus affects Coulomb energy and surface energy in such a way that the change in one and that in other tend to cancel one another partially. This is reasonable and to be expected, since the division of the nucleus increases (1) the separation between proton groups, thus reducing their Coulomb potential energy, (2) the total nuclear surface which increases the surface energy. Thus for spontaneous fission

$$-3.42\,A^{2/3} + 0.22\,Z^2/A^{1/3} \geq 0 \quad \text{or} \quad Z^2/A \geq 15 \qquad ...(14)$$

This relation shows that the fission should be energetically possible for nuclei with mass number A ≥ 85. However, the slow neutron fission does not take place even with many of the heavy nuclei. In order to explain this discrepancy Bohr and Wheeler considered the Coulomb's potential barrier of the two fragments at the instant of separation. The existence of this barrier prevents the immediate breaking of these two. If we denote the height of the Coulomb barrier by E_b, we can say that the nucleus will be unstable and break apart into two fragments if $E_f > E_b$. The barrier height corresponding to the Coulomb potential between the two symmetric fragments when they are just in contact with each other is given by

$$E_b = \left(\frac{1}{2}Z\right)^2 \frac{e^2}{A\pi \epsilon_0 (2R)} = \frac{Z^2 e^2}{32\pi \epsilon_0 R_0}\left(\frac{1}{2}A\right)^{1/2}$$

$$= 0.15\ Z^2/A^{1/3}\ \text{MeV} \qquad ...(15)$$

$$\therefore \quad E_b - E_f = 0.15\ Z^2/A^{1/3} - [-3.42\ A^{2/3} + 0.22\ Z^2/A^{1/3}]$$

$$= 3.42\ A^{2/3} - 0.07\ Z^2/A^{1/3} \qquad (16)$$

Thus the condition for stability gives

$$E_b - E_f \geq 0 \text{ or } Z^2/A \leq 49 \qquad ...(17)$$

For a particular nucleus, the closer the value of Z^2/A to 50, the shorter should be the half life for spontaneous fission. It is clear that Z^2/A may have value 50 for a nucleus of mass number 250, hence a nuclide (A > 250) would be too unstable to exist for more than 10^{-12} sec or less.

Computing E_b and E_f for various values of A for symmetric fission And plotting on the same graph, we get Fig. 9. The graph shows that for A $\simeq$ 250. E_b, becomes equal to E_f and E_f > Eb for nuclei with A > 250. This indicates that we do not expect nuclei with A > 250 to be found in nature.

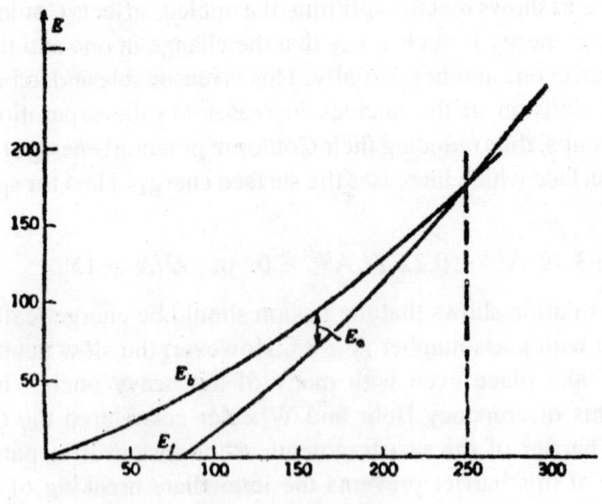

Fig. 9 : Variation of E_b and E_f with A.

Tins graph also shows that fission begins to become exoergic in the neighborhood of A $\simeq$ 85 in agreement with the earlier occurring result.

Bohr and Wheeler's Theory of Nuclear Fission

The first thorough theoretical treatment of this process was earned out by Bohr and Wheeler in 1939.

They applied a simple form of analysis (Legendre polynomial expansion) to express the radius r making angle θ with the axis of maximum deformation

$$r = R\left[1 + \sum_{l=0}^{\infty} \alpha_l P_t(\cos\theta)\right]$$

$$= R\,[1 + \alpha_2 P_2 (\cos\theta) + \alpha_3 P_3 (\cos\theta) + ...] \qquad ...(18)$$

where R is the radius of the spherical nucleus α_2, α_3, are the deformation parameters. Here $\alpha_0 = \alpha_1 = 0$, as the centre of mass of the drop is assumed to remain unchanged.

The surface energy of a spherical drop $E_s^{phere} = 4\pi R^2 T$, where T is the surface tension. Hence surface energy of the deformed drop

$$E_s = 4\pi R_0^2 A^{2/3} T\left[1 + \alpha_2\left(\frac{3}{2}\cos^2\theta - \frac{1}{2}\right) + ...\right]^2$$

$$= 4\pi R_0^2 A^{2/3} T\left[1 + \frac{2}{5}\alpha_2^2 + ...\right]$$

$$\therefore \quad \Delta E_s = E_s^{sphere}\left[\frac{2}{5}\alpha_2^2 + ...\right] \qquad ...(19)$$

The Coulomb energy of a spherical drop $E_C^{sphere} = \frac{3}{5} Z^2 e^2 / 4\pi \epsilon_0 R$

hence that of the deformed drop

$$E_e = \frac{3}{5}\frac{Z^2 e^2}{4\pi \epsilon_0 A^{1/3} R_0}\left[1 + \alpha_2\left(\frac{3}{2}\cos^2\theta - \frac{1}{2}\right) + ...\right]^{-1}$$

$$\therefore \Delta E_e = E_c^{sphere}\left[-\frac{1}{5}\alpha_2^2 - ...\right] = -E_c^{sphere}\left[\frac{1}{5}\alpha_2^2 + ...\right] \qquad ...(20)$$

Thus the total energy variation

$$\Delta E = \Delta E_s + \Delta E_c = \frac{1}{5}\alpha_2^2\left[2 E_c^{sphere} - E_c^{sphere}\right] \qquad ...(21)$$

If it is +ve, i.e., $2E_s^{sphere} > E_c^{sphere}$, the drop is stable to small distortions. Fissions may occur spontaneously if ΔE = –ve or

$$E_s^{sphere} < \frac{1}{2} E_c^{sphere}$$

$$\therefore \quad 4\pi R_0^2 A^{2/3} T < 3Z^2e^2/40\pi\varepsilon_0 A^{1/3} R_0 \text{ or } Z^2/A > 45$$

The ratio $\frac{E_c^{sphere}}{2E_s^{sphere}}$ known as *critical parameter*, represented by χ.

Thus when $\chi < 1$, the nuclear is stable against spontaneous fission. It is possible to estimate the degree of distortion of a nucleus in the critical state by equating the *critical or threshold* energy E_{th} to the total energy variation ΔE. From semi-empirical data $4\pi R_0^2 T$ = 13 MeV, hence for U^{238}, E_s^{sphere} = 520 MeV and E_c^{sphere} = 830 MeV and thus $\alpha_2^2 = 1/7$.

The energy that has to be imparted to the nucleus in order to reach this critical shape, the threshold energy is given as

$$E_{th} = 4\pi R^2 T\, f(\chi) = 4\pi R_0^2 A^{2/3} T\, f(\chi) = 17.8\ A2/3\ f(\chi) \text{ Mev} \quad ...(22)$$

This energy can be calculated by neglecting the second order change in energy due to the neck joining the two fragments

$$E_{th} = 2\left(4\pi R_0^2\right)T\left(\frac{1}{2}A\right)^{2/3} - 4\pi R_0^2 A^{2/3} T + 2\times\frac{3}{5}\times\left(\frac{1}{2}Ze\right)^2 /$$

$$4\pi \epsilon_0 R_0\left(\frac{1}{2}A\right)^{1/3} + \left(\frac{1}{2}Ze\right)^2 /8\pi \epsilon_0 R_0\left(\frac{1}{2}A\right)^{1/3} - \frac{3}{5}(Ze)^2/4\pi \epsilon_0 R_0 A^{1/3}$$

$$\text{or} \quad E_{th}/4\pi R_0^2 T A^{2/3} = 0.260 - 0.215\,\chi \quad ...(23)$$

For an uncharged droplet $\chi = 0$ and $f(0) = 0.260$, hence the critical energy is just the work done against surface tension in separating into two drops. For $\chi \approx 1$, a small deformation from the spherical shape causes the drop to reach the critical shape and separate.

If the critical energy is compared with the excitation energy, it becomes possible to predict fission probability. The excitation energy E_e, contributed to the resultant compound nucleus by the capture of a neutron, is equal to the binding energy of neutron in the compound nucleus and can be calculated by the relation.

$$E_e = B(A + 1, Z) - B(A, Z) = {}_ZM^A + M_n - {}_ZM^{A+1}$$

The values of the excitation energy calculated in this way for a number of heavy nuclei are listed in the table 13.2 and compared with the corresponding values of the critical energy. In reviewing the results, it is seen that for uranium—238 a critical deformation energy of 6.5 MeV is necessary for fission but it acquires only 5.9 MeV when it takes up a neutron of zero K. E. Thus no fission is possible with thermal neutrons with 0.03 eV energy. If the neutrons have a K. E. of 0.6 MeV fission becomes possible. Experiments indicate that neutrons of about 1 MeV energy are required. The fission cross section increases rapidly with neutron energy.

The situation is quite different with U^{235}. Here the excitation energy or the energy available by the capture of a slow neutron is greater than the threshold energy. It is evident that in this case thermal neutrons should be capable of causing fission of U^{235} nucleus. The reason for the difference in the excitation energy for U^{236} and U^{239} lies in the pairing term which appears in the semi-empirical mass formula.

The term makes a positive contribution of about 0.5 MeV to the binding energy of U^{236} but it is zero for U^{235} On the other hand the contribution is zero for U^{239} but is about 0.5 MeV for the U^{238}. Thus the odd-even effect is responsible for approximately 1 MeV of the difference in the energies gained by these isotopes upon the addition of a neutron.

Table 6

Compound Nucleus	E_e (MeV)	E_{th} (MeV)	$E_e - E_{th}$ (MeV)
Pa^{232}	5.4	5.0	0.4
Th^{233}	5.1	6.5	–1.4
U^{236}	6.6	5.5	1.1
Np^{238}	6.0	4.2	1.8
U^{239}	5.9	6.5	–0.6
Pu^{240}	6.4	.4.0	2.4

A general review of the changes in various types of nuclei, after neutron capture, shows that the liberation of energy is greater or the nuclei are likely to undergo fission with slow neutrons if the original nucleus contains an even number of protons and an odd number of neutrons or an odd numbers of both.

Whereas fast neutrons would be required for the nuclei containing odd-even or even-even numbers in the same mass region.

Quantum Effects

The values shown in table 13.2 do not agree well with measured values. This disagreement may be the result of two quantum mechanical effects.

(1) The fission may take place for excitation energies below the threshold due to the tunneling effect.

(2) The vibration of the drop in the distorted mode will have a zero point energy.

The tunneling effect is the origin of spontaneous fission. The fission problem is more difficult than the α-decay problem as it is very difficult to have a clear picture of the exact shape of the fission potential barrier. The fission barrier penetration probability

$$P \propto \exp\left[-\frac{2}{\hbar}\int_0^b \{2M(V-E)\}^{1/2}\, dr\right], \qquad ...(24)$$

where M is the reduced mass of the two fragment system, (V – E) is the –ve K. E. in the barrier of width b. For simplicity, let the barrier be of a parabolic form, given as

$$V = 1/2\ K\,(r - R)^2 \qquad ...(25)$$

where r is the separation of the two fission fragments and R is the separation at the top of the barrier. The width of the barrier b is given by the relation

$$E = 1/2\ K(b\ 1/2b)^2 \quad \text{or} \quad 1/2\ b = \sqrt{(2E/K)} \qquad ...(26)$$

Thus using parabolic function, we get

$$P \propto \exp\left[-\left(\frac{2\pi}{\hbar}\right)\left(\frac{M}{K}\right)^{1/2} E\right]$$

$$\propto \exp\left[-\left(\frac{b\pi}{2\hbar}\right)(2ME)^{1/2}\right] \qquad ...(27)$$

For U^{238}, E ~ 6 MeV, b = 1.5 × 10^{-14} m and M = 240/4 = 60 mu

$$\therefore \quad P \propto \exp(-100) \qquad ...(28)$$

Frankel and Metropolis obtained an expression for the life time for spontaneous fission using the idea of barrier penetration as

$$t = 10^{-21} \times 10^{7.85}{}_{th}\ \text{sec} \qquad ...(29)$$

Scaborg calculated E_{th} for various heavy elements and found that the fission rate could be determined by the formula

$$t = 10^{178-3.75(Z^2/A)} \text{ sec} \quad ...(30)$$

After comparing with eqn (29), he obtained the relation

$$E_{th} = 19.0 - 0.36\ Z^2/A \quad ...(31)$$

which is in close agreement with experiment.

GENERAL ASPECT OF REACTOR DESIGN

In spite of numerous possible variations in the design and components of reactor systems, there are number of general features which all reactors possess in common to a greater or lesser extent Let us consider and compare few features in the given order.

Reactor Coolants

The materials employed, to remove the heat that is being generated in the reactor core as a result of fissioing taking place, are known as *coolants*. These materials are circulated through the core tor the purpose of abstracting the heat and transferring it to the outside of the core. An ideal coolant should have as little effect on the neutron as possible, ideally it should not absorb nor moderate the neutrons, should not react chemically with the other materials which it contacts at the system, should not break up under irradiation, should acquire intense long lived radioactivity during its passages through the reactor, should have a low vapour pressure at the operating temperature of the reactor, should be reasonably easy to handle, should be able to remove large amounts of heat for a small expenditure of pumping power and should not be costly. Actually no one coolant meets all these requirements, a final choice for any particular reactor is a compromise based on these requirements. The materials proposed for coolants are ordinary water, heavy water, liquid metals such as sodium or sodium potassium alloy or mercury, organic liquids and gases. Each type has its advantages and disadvantages.

At ordinary temperatures both heavy and light water are good coolants. For reactors operating at high temperatures, high pressures are required to prevent boiling. Liquid metals have been proposed for use at higher temperatures. The best of these is sodium. It does not require pressurization even at very high temperatures. On the other hand, it is very reactive with water and oxygen and becomes radioactive due to neutron capture when passed through reactor. As a compromise between water and liquid sodium, certain organic compounds e.g. polyphenyls, were suggested. However the heat removal properties of these hydrocarbons are inferior to those of water or liquid sodium. In several of the earliest reactors air was used as a coolant. It is not satisfactory at high temperatures because it reacts chemically with so many materials. Carbon dioxide is also employed in large power reactors. Since it is used as coolant in graphite-moderated reactors. The chemical reaction with graphite is a problem at temperatures greater than 500°C. By the addition of some carbon-monoxide or methane it can be used upto 600°C. Actually helium is a good coolant but it is very costly.

Control Materials

In thermal reactors control is achieved by means of a neutron absorbing materials. The material should not become radioactive as a result of neutron capture. Became of its availability and ease of fabrication cadmium was used in early reactors. It can be utilized at low temperature work only because it has low melting point. An alloy of silver with 15 per cent of indium and 5 percent of cadmium has been also employed in some reactors. This alloy has a higher melting point than Cd and large cross-section for neutron capture also over a wider energy range. The most common neutron absorber used for reactor control is boron. This element has very high melting point and large cross-sections for neutron absorption-Boron is often used incorporation with stainless steel, aluminium or carbon.

The control elements are commonly located in the core in the form of either rods or plates, but in some reactors it is more convenient to have the control elements in the reflector close to the core. In a thermal reactor the control rods are moved in to decrease the fission rate or neutron flux and out to increase it. In fast reactors materials cannot be used because the cross-sections are small. In this case control is achieved by adjusting neutron escape by moving part of the reflector or some of the fuel elements.

Reactor Shielding

All nuclear reactors, except those operating at very low powers, are sources of intense neutron and γ-radiation and therefore, represent hazard to persons in the immediate vicinity of the reactor. Provisions for their health protection are made by surrounding the reactor core with a radiation shield. This shield is generally known as the *biological shield* because its primary function is of health protection. It consists generally a layer of concrete, about 6 to 8 ft. thick and is capable of absorbing both γ-rays and neutrons. In reactors operating at high powers, that part of the shield which is in immediate contact with the core heats up considerably and requires special cooling facilities in order to prevent it from cracking or suffering other heat damage. Shield required for this purpose is known as thermal shield. It is fairly close to the core and consists of a few inches of iron or steel.

Fuel

The material containing the fissile isotope is called the reactor fuel. The fuels that can be used are uranium containing U^{235} in its natural concentration of 0.715% or in an enriched proportion. Th^{232} and U^{235}, which can be converted partly at least into fissile material. The form in which a fuel is used in a reactor depends upon various circumstances. Solid fuel elements are used in most of the reactors. Some reactors have been designed which make use of fluid fuels.

Moderators and Reflectors

Materials used to reduce the neutron energy, known an moderators, are graphite, light water, heavy water, beryllium and its oxide and possibly certain organic compounds. As previously noted, the good moderators are usually materials of low mass number having small absorption cross-section and large slowing down power.

The ordinary water has the greatest *sdp* but an appreciable neutron absorption cross-section. It is attractive as a moderator because of its low cost. Heavy water is the best available material for neutron moderation. It has a very small absorption cross-section and the highest moderating ratio of all moderators.

Carbon in the form of graphite has been employed as moderator or reflector in a number of reactors. It has a very small absorption cross-section and the second highest moderating ratio of all moderators. Both beryllium and its oxide are good moderators having small absorption cross-section and the best sdp of all metals. Metallic beryllium is much more expensive than other moderators. Hydrogeneous organic compounds have proved to be suitable as moderators. Above mentioned materials arc not suitable for use as the reflector for the reactor in which fast neutrons are used. To avoid this situation the reflector must be a material of high mass number.

SOLVED EXAMPLES

Example 1:

Calculate the critical volume of a spherical thermal reactor that uses U^{235} as fuel and heavy water as moderator in molecule ratio of $1 : 10^5$.

Solution:

$$\text{Thermal utilization factor } f = \frac{\sigma_a(U)}{\sigma_a(U) + \sigma_a(M)\dfrac{N_m}{N_u}}$$

$$= \frac{698}{698 + 0.00092 \times 10^5} = 0.8948$$

$$\therefore k_\infty \eta f = 2.08 \times 0.8948 = 1.819$$

Thermal diffusion length for the mixture

$$L^2 = L_m{}^2 (1 - f) = 1.70^2 (1 - 0.8948) = 0.3038 \text{ m}^2$$

$$\therefore \quad k_{eff} = \frac{k_\infty e^{-B^2}}{1 + L^2B^2} = \frac{1.819\, e^{-0.125B^2}}{1 + 0.3038\, B^2} = 1$$

or $B^2 = 2.192 \text{ m}^2$

$\therefore$ Critical volume $V_C = 120/53 = 37 \text{ m}^3$.

Example 2:

What would be the length of she side of a cubical reactor having geometrical buckling 65. If the reactor is spherical, calculate its critical radius.

Solution:

$$\text{Geometrical buckling } B^2 = \frac{\pi^2}{a^2} + \frac{\pi^2}{b^2} + \frac{\pi^2}{c^2}$$

For a cubical reactor a = b = c

$$\therefore\ B^2 = \frac{3\pi^2}{a^2} \text{ or } a^2 = \frac{3\pi^2}{B^2} = 0.46154$$

$\therefore$ a = 0.674 m = 67.4 cm

Critical radius of a spherical reactor

R = π/B = 3.141/8 = 0.39 m = 39 cm.

Example 3:

A thermonuclear device consists of a torus of diameter 3 m with a tube of diameter 1 m, containing deuterium gas at 10^{-2} mm mercury pressure and at room temperature. A bank of capacitors of 1200 μf is dencharged through the tube at 40 kV. If only 70% of the electrical energy is transformed to plasma kinetic energy, what is the maximum temperature attained ? Assuming that the energy is equally shared between the deuterons and electrons in the plasma.

Solution:

Cross-sectional area of the torus 1/4π meter^2

Circumference = 3π meter.

$$\therefore \text{ Volume of Torus } = \frac{3\pi^2}{4}\ 7.4\ m^3$$

Pressure of the gas $10^{-5} \times 13.6 \times 10^3 \times 9.81 = 1.34$ N/m^2.

From gas eqn. PV = NkT, we have Nk = 1.34 × 7.4/293 = 0.0338.

The energy obtained from the discharge

$= 1/2\ CV^2 = 1/2 \times 1200 \times 10^{-6} (4 \times 10^4)^2 = 9.6 \times 10^6$ joules.

$\therefore$ Energy transformed to plasma K.E. = 9.6×10^4 joules.

We know that the average kinetic energy associated with the gas molecule is given as $E_r = 3/2\ NKT_k$. Since each deuterium molecule produces two ions and two electrons, hence

$$4\,\frac{3}{2}\,NkT_k = 9.6 \times 10^4 \text{ or } T_k = 4.75 \times 10^5\ °K$$

Example 4:

Calculate the amount of energy released if all the deuterium atoms in the water in lake of area about 10^5 sq. miles and of depth $1/20^{th}$ mile are used up in fission,

Solution:

The volume of water = $10^5 \times 1/20 = 5000$ cubic miles.

$\therefore$ Mass of water = $5000 \times (1.6 \times 10^3)^3 \times 10^3$ kg.

or No. of molecules of water = $2.13 \times 10^{16} \times 6.02 \times 10^{26}/18$

$= 7.13 \times 10^{41}$ molecules.

As the abundance of deuterium is 0.0156% so that the total number of deuterium atoms = $7.13 \times 10^{41} \times 2 \times 0.0156 \times 10^{-2} = 2.22 \times 10^{38}$. M the fusion of 6 deuterium atoms gives an energy release of 43 MeV, hence the total energy released = $2.22 \times 10^{33} \times (43/6)$

Example 5:

Calculate k_∞ for a homogereous, natural uranium-heavy water moderated assembly which contain 50 molecules of moderator per molecule of uranium. Assume natural uranium to contain one part U^{235} to 139 parts of U^{238} and use the following constants : For uranium, $\sigma_a(U)$ = 7.68 barns, $\sigma_s(U)$ barns; and for D_2O, σ_a = 0.00092 barn, σ_s = 10.6 barns; and ξ = 0.570.

Solution:

Thermel utilization factor $f = \dfrac{N_u\sigma(U)}{N_u\sigma_a(U) + N_u\sigma_a(M)}$

$$= \frac{1}{1 + \dfrac{N_m\sigma_a(M)}{N_u\sigma_a(U)}} = \frac{1}{1 + \dfrac{0.00092}{7.68}} = 0.994$$

Number of U^{238} atoms only per unit volume $N_0 = N_u(139/140)$

$$\therefore \quad \frac{\Sigma_s}{N_0} = \frac{N_m}{N_0}\sigma_s(M) = N_m \cdot \frac{140}{139 N_u} \times \sigma_s(M)$$

$$= \frac{N_m}{N_u} \times \frac{140}{139} \times \sigma_s(M) = 50 \times \frac{140}{139} \times 10.6 = 530 \text{ barns}$$

$$\therefore \quad \int (\sigma_s)_{eff} \frac{dE}{E} = 3.85\left(\frac{\Sigma_s}{N_0}\right)^{0.415} = 60.2 \text{ barns}$$

$\therefore$ Resonance escape probability $P = e^{-60.2/530 \times 0.57} = 0.820$

Hence $k_\infty = \epsilon\eta p f = 1 \times 1.34 \times 0.820 \times 0.994 = 1.09$

This assembly is, therefore, not capable of sustaining a chain reaction.

Example 6:

Estimate the relative loss due to leakage for a critical thermal reactor employing U^{235} and graphite in on atom ratio of 1: 705. Given : For uranium $\sigma_a = 0.003$ barn, $L_m = 54$ cm, $\tau_0 = 364$ cm^2. For uranium $\sigma_a = 698$ barns and $\gamma_1 = 2.08$. Also given $p = 1$ and $\in = 1$.

Solution:

Since p = 1 and ∈ = 1, hence $k_\infty = \eta f$.

$$\text{Thermal utilization factor } f = \frac{N_{0(235)}\ \sigma_{a(235)}}{N_{0(235)}\ \sigma_{a(235)} + N_{0(c)}\ \sigma_{a(c)}}$$

$$= \frac{1}{1 + \dfrac{N_{0(c)}\ \sigma_{a(c)}}{N_{0(235)}\ \sigma_{a(235)}}} = \frac{1}{1 + \dfrac{10^5 \times 0.003}{698}}$$

$= 0.6995.$

$\therefore \quad k_\infty = 2.08 \times 0.6995 = 1.454.$

Thermal diffusion length for the mixture $L^2 = L_m^{\ 2}\ (I - f)$

$= (0.54)^2\ (1 - 0.6995) = 8.76 \times 10^{-2}$ m^2

We know that for critical reactor

$$k_{eff} = \frac{k_\infty\, e^{-B^2\tau}}{1 + L^2B^2} = \frac{1.454\ e^{-364 \times 10^{-4} B^2}}{1 + 8.76 \times 10^{-2}\ B^2} = 1$$

$\therefore \quad B^2 = 3.27$. Substituting this value of B^2, we get fast non-leakage probability $l_f = e^{-B^2\tau}$

$\therefore$ Fast leakage probability $= 1 - l_f = 1.112 = 11.2\%$.

Thermal non-leakage probability $l_{th} = 1/(1 + L^2B^2) = 0.778$.

$\therefore$ Thermal leakage probability $1 - l_{th} = 1 - 0.778 = 0.222$

Since only a fraction of 0.888 of the original neutrons have reached thermal energies. Hence the fraction of original number lost by thermal leakage, is given by

$$0.222 \times 0.888 = 0.197 = 19.7\%$$

$\therefore$ Total leakage factor = 11.2% + 19.7% = 30.9%.

It can also be obtained by other method.

$$\text{Total non-leakage factor } l_f\, l_{th} = \frac{k_{eff}}{k_\infty} = \frac{1}{1.464} = 0.688$$

∴ Total leakage factor = 1 – 0.688 = 0.312 = 31.2%.

Example 7:

Calculate the fission rate for U^{235} required to produce 2 wait and the amount of energy that is released In the complete fissioning of 1/2 kg of U^{235}.

Solution:

As we know that 200 MeV energy is released per fission of U^{235}

∴ Fission rate = 2 watt/200 MeV per fission

$= 6.15 \times 10^{10}$ fission/sec.

No. of U^{235} nuclei in 1/2 kg. of U^{235} = (0.5/235) × 6.0247 × 1026

On fissioning this number of U^{235} nuclei, the energy release will be $= (0.5/235) \times 6.0247 \times 10^{26} \times 200$ MeV

$= 2.57 \times 10^{26}$ MeV $= 10^{10}$ kilocalories.

Example 8:

Calculate the excitation energy for U^{236*} and for U^{239*}. Estimate the rare of spontaneous fissioning of 1 gm of U^{235}, given its half life ~3 × 70^{17} years.

Solution:

The excitation energy $E_e = B\ (Z, A + 1) - B(Z, A)$.

For U^{236}

$$E_e = a_v(236 - 235) - a_s(236^{2/3} - 235^{2/3}) - a_c(92^2/236^{1/2} - 92^2/235^{1/2})$$
$$-a_a\ (52^2/236 - 51^2/235) + a_v(236^{-3/4} - 0)$$
$$= a_v + - 0.1\ a_s - 22\ a_c - 0.4\ a_a + 0.0167\ a_p = 6.8 \text{ MeV}$$

Similarly for U^{239}, $E_e = 5.9$ MeV.

Rate of spontaneous fissioning $\dfrac{dN}{dt} = \lambda N$, where

$N = 6.023 \times 10^{23}/235$ and $\lambda = \log_e 2/T_{1/2}$

$= 0.693/3 \times 10^{17} \times 3.15 \times 10^7$

$\therefore \quad dN/dt = 2 \times 10^{-4}\ sec^{-1} = 0.7\ hr^{-1}$.

Example 9:

Calculate η for thermal neutron induced, fission of a uranium mixture containing U^{235} and U^{238} isotopes in a 1 : 20 ratio.

Given $\nu = 2.43$, $\sigma_a(U^{235}) = 683$, $\sigma_a(U^{238}) = 2.73b$, $\sigma_f(U^{235}) = 583b$

Solution:

Average number of fission neutrons released per absorption

$$\eta = \nu(\sigma_a/\sigma_a)$$

For a mixture

$$\sigma_a = \frac{N_0(235)\,\sigma_a(235) + N_0(238)\sigma_a(238)}{N_0\,(235) + N_0(238)} = \frac{683 + 20 \times 2.73}{1 + 20}$$

$= 35.1$ barns.

$$\text{and } \sigma_f = \frac{N_0(235)\,\sigma_f(235)}{N_0\,(235) + N_0(238)} = \frac{283}{1 + 20} = 27.8 \text{ barns.}$$

Hence $\eta = 2.43 \times (27.8/35.1) = 1.924$.

Example 10:

The world's first water boiler reactor, the LOPO nuclear reactor, used an enriched uranium sulphate solution made up as follows:

Element	*weight*	*σ_a(barn)*	*σ_s(barn)*
U^{235}	*580 gm*	*698*	*10*
U^{238}	*3378 gm*	*2.75*	*8.3*
S	*534 gm*	*0.49*	*LI*
0	*14065 gm*	*0.0002*	*4.2*
H	*1573 gm*	*0.33*	*20*

Calculate k_∞ for this reactor.

Solution:

Thermal utilization factor

$$f = \frac{1}{1 + \frac{N_{(238)}\sigma_{a(238)}}{N_{(235)}\sigma_{a(235)}} + \frac{N_{(s)}\sigma_{a(s)}}{N_{(235)}\sigma_{a(235)}} + \frac{N_{(0)}\sigma_{a(0)}}{N_{(235)}\sigma_{a(235)}} + \frac{N_{(H)}\sigma_{(H)}}{N_{(235)}\sigma_{a(235)}}}$$

$$f = \frac{1}{1 + 5.75 \times 3.94 \times 10^{-3} \times 6.76 \times 7.15 \times 10^{-4}}$$

$$+356 \times 2.86 \times 10^{-5} + 637 \times 4.73 \times 10^{-4}$$

$$= 0.754$$

$$\frac{\Sigma_s}{N_0} = \frac{N_{(0)}}{N_{(238)}}\sigma_{s(0)} + \frac{N_{(H)}}{N_{(238)}}\sigma_{s(H)} = \frac{879.25}{14.19} \times 4.2 + \frac{1573}{14.19} \times 20 = 2475 \text{ barns}$$

$$\therefore \quad \int_E^{E_0} (\sigma_a)_{eff} \frac{dE}{E} = 3.85 \left(\frac{\Sigma_s}{N_0}\right)^{0.415} = 99 \text{ barns}$$

$$\text{and} \quad \frac{N_0}{\xi\Sigma_s} = \frac{N_0}{\xi_{(0)}\sigma_{s(0)} N_{(0)} + \xi_{(H)}\sigma_{(H)} N_{(H)}}$$

$$= \frac{14.19}{0.120 \times 4.2 \times 879 + 1 \times 120 \times 1573} = \frac{1}{2249}$$

$$\therefore \quad p = e^{-99/2249} = 0.957$$

Example 11:

Calculate k_∞ for an enriched uranium-graphite moderated assembly which contains 400 molecules of moderator per molecule of uranium and a U^{238}/U^{235} ratio of 75. Given: For graphite, σ_a = 0.0032 barn, σ_s = 4.8 barns. ξ = 0.758; For uranium, $\sigma_f(235)$ = 590, $\sigma_a(235)$ = 698, $\sigma_a(238)$ = 2.75 and ν = 2.46.

Solution:

Let us first calculate absorption cross section $\sigma_a(U)$ and number of fission neutrons per neutron absorbed η_u for the enriched fuel.

$$\sigma_a(U) = [N(235)\,\sigma_a(235) + N(238)\,\sigma_a(238)]/[N(235) + N(238)]$$

$$= \frac{\sigma_a(235) + \sigma_a(238)\,N(238)/N(235)}{1 + N(238)/N(235)} = \frac{698 + 2.75 \times 75}{1 + 75}$$

$$= 11.89 \text{ barns.}$$

$$\eta_u = [N(235)\ \sigma_f(235)\ v]/[N(235)\ \sigma_a(235) + N(238)\ \sigma_a(238)]$$

$$= \frac{v\sigma_f(235)}{\sigma_a(235) + \sigma_a(238)\ N(238)/N(235)} = \frac{2.46 \times 590}{698 + 2.75 \times 75} = 1.60$$

Calculate f and p and then k_∞ as we did, in the last example.

This value of p is considerably greater then the values obtained for natural uranium. Since f is evaluated with respect to U^{235}, the matching value of η to be used now is 2.08. Hence

$$k_\infty = \epsilon\eta\ pf = 1 \times 2.08 \times 0.957 \times 0.754 = 1.5.$$

EXERCISES

1. The rate of energy delivery to the earth from the sun, 1.5×10^8 km distant, is known as the solar constant, with a value of 2.0 cal cm^{-1} min^{-1}. Calculate the rate at which -the sun is losing mass to supply this radiated energy. What fraction of its mass of 2 $\times 10^{33}$ gm does the sun lose per year?
2. Calculate the thermal utilization factor for an aqueous fuel solution containing 10% by weight, of U^{235}.
3. Calculate p for a homogeneous, natural uranium-BeO assembly with molecular ratio of 1 : 200.
4. Calculate k_∞ for a homogeneous, natural uranium-graphite-moderated assembly with contains 300 molecules of graphite per molecule of uranium. Assume natural uranium to contain one part of U^{235} to 139 parts of U^{235}. Given; For natural Uranium, $\sigma_a = 7.68$ barns, $\sigma_s = 8.3$ barns; For graphite, $\sigma_a = 0.0032$ barn, $\sigma_s = 4.8$ barn, and $\xi = 0.158$.
5. Calculate k_∞ lot an enriched uranium-graphite-moderated assembly, using 400 molecules of graphite to 1 molecule of uranium and a U^{238}/U^{235} ratio of 70.
6. A power reactor operating at a power level of 10^5 watts goes prompt critical due a failure of the control rod system. Assume a multiplying factor of 1.09, and that the negative temperature coefficient stopped the reaction in 5 m sec. Estimate the thermal energy generated during the runaway.
7. A critical reactor of cubical shape uses U^{235} and ordinary water in a molar ratio of 1:400. Calculate the critical mass of U^{235} for this reactor.

8. In a neutron induced fission of U^{235} two stable end products Mo^{98} and Xe^{136} are often found. Assuming that these isotopes have come from the original fission process, find (a) what elementary particles are released, (b) the mass defect of the reaction and (c) the equivalent energy released.

9. In the fission reaction $Sn^{118} + p \rightarrow Na^{24} + Zr^{94} + n$, determine (a) the energy equivalent of the mass difference between reactants and products, (b) the excitation energy necessary for passage over the potential barrier and (c) the threshold energy for the reaction.

10. How much U^{235} and U^{238} will be lost in 1 year from spontaneous fission in a stockpile of 1000 kg of natural uranium? How much will be lost from α-decay?

11. Liquid sodium used as a reactor coolant has a cross-section for thermal activation of 600 mb. The circulating sodium spends about 1/4 of its time in the reactor core, where the thermal flux is 8.6×10^{12} n cm^{-2} sec^{-1}. What is the specific activity of the coolant at equilibrium? How long will be required for the coolant to reach 3/4 of equilibrium activity?

12. The total volume of water in the oceans is 1.34×10^9. What is the total mass of deuterium contained in the water? What will be the total energy available from allot this deuterium through reaction $H^2 - H^2 \rightarrow H^3 + p + 4.04$ MeV? How much mass will be lost in this energy release?

3
Nuclear Particles

INTRODUCTION

If anti-matter exists it would then be made up of anti-electrons, i.e. positions, anti-protons and neutrons. Thus we see that seven particles could explain both matter and anti-matter. In 1935, Yukawa postulated the existence of another particle, with a mass m $\simeq$ 200 m_e *as the field particle for the strong nuclear forces*. Recently the extensive studies made partly on high energycosmic ray particles and even more, with the help of high energy accelerators have revealed the existence of numerous new nuclear particles.

Apart from a dozen or so the particles have very short lifetimes, very much less than 10^{-6} sec. In 1932, when Chadwick identified the neutron and Heisenberg suggested that atomic nuclei consisted of neutrons and protons, it seemed as if p, n and e^- were sufficient to account for the structure of matter. Besides these there was the photon, *the intermediary* or *field particle for electromagnetic forces*, such as exist between the nucleus and electrons in the atom.

They cannot therefore be regarded as normal constituents of matter. They arc characterised by the parameters, miss, spin, electric charge and magnetic moment They have been described by such adjectives, as *fundamental, strange* and *elementary*, but none of these is quite appropriate.

The word fundamental implies that the particles are the basic building blocks of matter, but unstability of most of the particles indicates that the great majority are certainly not. It is true that their behaviour was strange in the early 1950, but it is much less now. For the want of better one the term elementary particles is now commonly used. These particles are elementary in much the same sense as are the chemical elements.

CLASSIFICATION OF ELEMENTARY PARTICLES

The elementary particles are separated into two general groups, called *bosons* and *fermions*. These two groups have different types of spin and their behaviour is controlled respectively by a different kind of statistic (i.e. the Bose statistic or the Fermi statistic, hence the names). Bosons are particles with intrinsic angular momentum equal to an integral multiple of $\hbar$. Fermions are all those particles in which the spin is half integral. The most important difference between the two classes of particles is that *there is no conservation law controlling the total number of bosons in the Universe, whereas the total number of fermions is strictly conserved.*

Boson is a term, which not only includes material particles but also includes those quanta and photons which arise from interactions. Thus in the case of the simple electromagnetic field the bosons are merely the light photons or the X-ray photons. The *photon* has a mass of zero and a spin of unity and is consequently described as a massless boson. A massless boson, called a *graviton* with a probable spin of two units has been postulated as a field particle for gravity. These bosons, created by the electromagnetic field, are essentially of one kind, while the bosons formed in the strong interaction are of two distinct kinds. First there are those which are known as *pions or π-mesons* (π^+, π^- and π^o). The second group of bosons are much heavier than that of pions, and are known is *kaons or K-mesons* (K^+, K^- and K^o).

The fermions fall in two main classes, according to whether they are lighter than mesons, or heavier. Those in the lighter group are often called *leptons (after the Greek word meaning light in weight)*, whilst those in the heavier group are called *baryons (after the Greek lord for heavy)*. The leptons are the *electrons, muons* and *neutrinos* and *their anti-particles*. These arc all with masses less than the pions and with spin half. Leptons interact weakly with other particles. The total number of leptons minus the total number of anti-leptons remains unchanged in all reactions and decay processes involving leptons and anti-leptons. The baryons consist of the two nucleons with their *anil-particles* (n^o, $\overline{n^o}$; $\bullet^+$, p^-) and the *hyperons*. Hyperons are the extremely unstable somewhat heavier particles and can be divided into four sub-groups, Λ^o-particle (a neutral particle of mass about 180 m_e), the Σ-particlcs (Σ^-, Σ^o and Σ^+ with masses in the range 2320 to 2340 m_e), the Ξ-particles (Ξ^- and Ξ^o with masses near 2580 m_e) and the Ω^--particle (of mass about

3284 m_e). There is no reason to doubt the existence of the anti-particles of these fermions. The total number of baryons minus the total number of anti-baryons is absolutely conserved in all interactions.

RESONANCE STATES

Most resonant or unstable states result from the strong interaction of a high energy meson (π or K) with a proton. Sometimes, resonant states also result from the interaction with a neutron or a deuteron. These states can be assigned mass, charge and spin consistent with the conservation laws. The earliest evidence for the formation of resonant states, also referred to as *resonance particles* or *resonances*, came in 1951 from a group working with Fermi at the Chicago University. In the course of studies of the scattering of positive and negative pions of various energies by protons, it was observed that the cross-sections increased rapidly at energies above about 60 MeV. In 1957, K.A. Brueckner at Indiana University suggested that a resonance interaction occurred between the proton and the pion, much as the resonance capture of a neutron by a nucleus was associated with a maximum in the cross-section. *It is for this reason that the term resonance or resonant state came into general use.* At the present time, the name is applied to a particle state of very short life (~ 10^{-23} sec) which decays into two or more particles. Because of their extremely short lifetimes, resonant states are described as unstable particles.

In spite of their extremely short lives, two general methods have been used for detecting resonances. *One is to measuse the scattering cross-section as a function of energy of the incident meson.* The plot of the results indicates the formation of a resonant state. The rest mass energy of this state can be calculated from the energy of the scattered particle corresponding to the maximum cross section. The average lifetime of the resonance, equal to the uncertainty in the time, $\tau = \Delta t = \hbar/\Delta E$, where ΔE is the uncertainty in energy which is equal to the full width of the peak at half the maximum height. Suppose the peak width at half maximum is found to be 100 MeV, the mean life of the resonant state is $6.6 \times 10^{-21}/100 = 6.6 \times 10^{-21}$ sec. The scattering cross-sections of positive and negative pious by protons are shown in Fig. 12. The upper curve shows that the maximum of the resonant peak is at a pion energy of 195 MeV in the lab system. On converting it in the proton pion centre of mass system the rest mass energy of the resonant state is found to be 1238 MeV. The full width at half maximum is 120 MeV, corresponding to the life time of 5×10^{-21} sec. This short life indicates decay by the strong interaction. This particular resonance state was attributed to a state with an isospin of 3/2 by K.A. Brueckner, and was originally represented by the symbol $N_{3/2}^{*}$ indicating a *nucleon excited state with* T = 3/2. It is now designed as Δ(1238). Theory predicts that in the formation of an T = 3/2 resonance the maximum scattering cross-section for π^{+}-p should be three times greater than for π^{-}-p. It is same as indicated by Fig. 1.

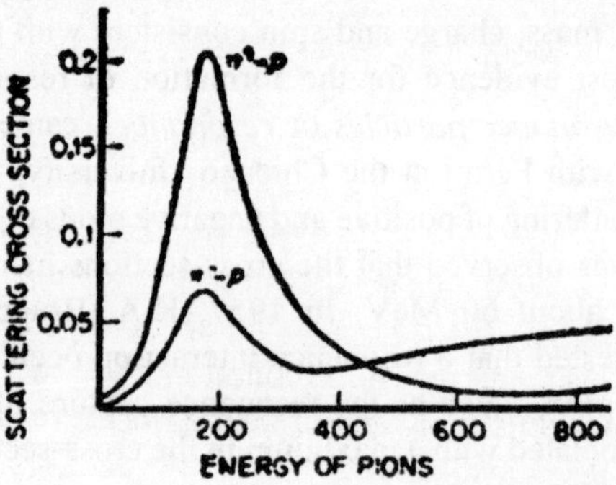

Fig. 1 : Scattering cross-ssctions of π^{+}, π^{-}-mesons by protons.

When the cross-sections for scattering of positive and negative pions with energies in excess of 300 MeV by protons are measured, several additional resonant states are observed. Some of these are more prominent in the π^{+} interactions, whereas other are in the π^{+}-interactions. These states can be assigned mass, charge and spin consistent with the conservation laws. The masses of some of well established resonant states are 1518, 1688, 1920 and 2360 MeV. These resonant states all have same mean lifetimes ~ 10^{-23} sec.

The second method for detecting resonances is based on a study of the energy distribution among the products, as determined from bubble chamber events.

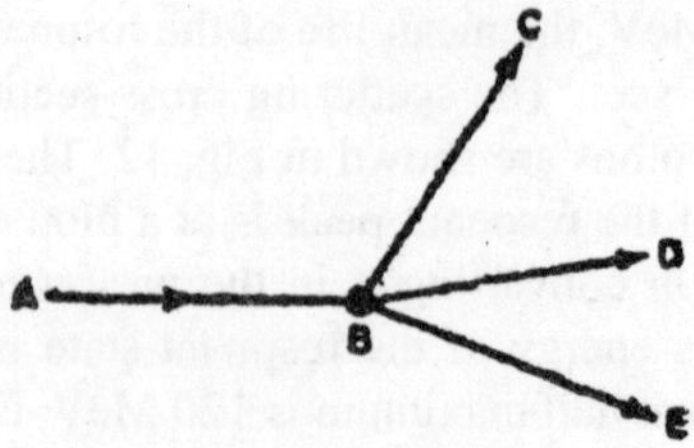

Fig. 2 : Bubble Chamber track.

Suppose a high energy particle A interacts with a stationary particle B, in the bubble chamber to produce a particle C plus a resonant state R. The latter decays into two particles D and E in such a very short time that it leaves no track in the chamber. If all these particles are charged, the tracks obtained in the bubble chamber will be as shown in Fig. 2. The problem is thus to distinguish between the postulated process.

$$A + B \rightarrow C + R \rightarrow C + D \rightarrow E \qquad ...(46)$$

and the direct reaction

$$A + B \rightarrow C + D + E. \qquad ...(47)$$

The problem can be solved by a kinematic analysis of the bubble chamber event made with the aid of a computer. The available energy will be distributed among the three particles in a random manner, as given by the law of conservation of momentum, in the case in which these particles are formed by a direct reaction. If large number of events of this type are examined, it will be found that the energy of any particle, say of C, has a spectrum of values ranging from zero to a maximum equal to the total energy available.

If we consider another case in which a resonant state R is first formed together with a particle C. The available energy will be divided between them in a definite manner which depends only on the relative masses of these two. Thus in a large number of events, there will be many cases in which particle C has a specific energy value and a peak will be observed in the energy distribution curve. Energy of the resonance R can be calculated from the energy of the particle C corresponding to this peak. The mean life-time of the resonant state can be calculated by knowing the uncertainty in energy which is equal to the width of the peak at half maximum.

The problem that which of the three emergent tracks in the bubble chamber event represents particle C was solved by a group at the University of California, Berkeley in 1960. In the search for a hyperon-pion resonance analogous to the proton-pion resonance described earlier, the reaction was

$$K^- + p^+ \rightarrow \Lambda^o + \pi^+ + \pi^-.$$

Many hundreds of plates were analyzed by a computer. Some of the results suggested that the law of conservation of linear momentum was being violated and two resulting particles were indicated rather than three. Possibilities were

$$K^- + p^+ \rightarrow Y^{*+} + \pi^-$$
$$K^- + p^+ \rightarrow Y^{*-} + \pi^+,$$

where Y* was suggested a new resonance particle showing strong decay in 10^{-33} sec into

$$Y^{*+} \rightarrow \Lambda^o + \pi^+$$

$$Y^{*-} \rightarrow \Lambda^o + \pi^-.$$

Thus we see that $\Lambda^o\pi^+$ and $\Lambda^o\pi^-$ resonances are formed, the rest mass corresponding to the pion energy of 280 MeV is calculated to be 1385 MeV. This resonance state, originally designed Y* is now preferably indicated by the symbol Σ^* (1385). The widths at half maximum of the energy peaks at 280 MeV are both 40 MeV. This shows that the Y* decays by the strong interaction. It is built from either $\Lambda^o\ \pi^+$ or $\Lambda^o\ \pi^-$, therefore, S = –1 and T = 1. Its baryon number is one and angular momentum is 3/2.

Evidence for an excited slate of a K-particle has been obtained from hydrogen bubble chamber events

$$K^+ + p \rightarrow K^{*+} + p;\ K^* \rightarrow \overline{K}^o + \pi$$
$$K^- + p \rightarrow K^{*-} + p$$
$$K^- + p \rightarrow \overline{K}^{*o} + m$$

The K* excited state has rest mass 892.4 MeV and isotopic spin half. It has a spin 1 and two values of strangeness S = ± 1 in the two charge states as for the K-meson. In addition to these resonances, large number of others have been observed.

Nambu in 1957 suggested the existence of a heavy neutral meson having zero isospin and unit angular momentum. Chew in 1960 had also shown that such a meson should exist on dynamical grounds as a 3π resonance, or bound state. The first experimental evidence for a three pion resonance, with T = 0 and I = 1, was found by a study of four pronged annihilation events when the 72 inch. Berkeley hydrogen bubble chamber was exposed to a high energy anti-proton beam. The over all process is

$$\bar{p} + p \rightarrow \pi^+ + \pi^- + \pi^o + \pi^+ + \pi^-.$$

In order to determine if a resonant state involving three pions are formed as an intermediate stage, three groups of pion triplets were studied, namely, singly charged $\pi^+\pi^+\pi^-$ and $\pi^-\pi^-\pi^+$, doubly charged $\pi^+\pi^-\pi^o$ and $\pi^-\pi^-\pi^o$ and neutral $\pi^+\pi^-\pi^o$. About 800 bubble chamber events were analyzed and for each of the three groups a histogram was obtained of the number of triplets against the effective mass of the triplet. For the charged triplets, the histograms were in good agreement with the phase space distribution. The maximum is observed only in the case of neutral triplet. It follows, therefore, that a neutral resonant state is formed and that it decays into positive, negative and neutral pion. This resonant state is known as the omega (ω)-meson. Thus the above reaction can be written as two stage process.

$$\bar{p} + p \rightarrow \omega^o + \pi^+ + \pi^-$$

$$\omega^o \rightarrow \pi^+ + \pi^- + \pi^o.$$

In a bubble chamber study of the four pronged proton-antiproton annihilation events gave the evidence for a three pion resonance ω^o-meson. The state with charge zero shows a marked peack at a mass of 787 MeV but no such peak appears in the states with charge 1 or 2. The favoured state, therefore, has T = 0, I = 1 and parity odd. Gelfand *et al.* (1963) observed mass of 784.0 ± 0.9 MeV, its decay into 3 π with width 9.5 ± 2.1 MeV and a life time of $(0.69 \pm 0.15) \times 10^{-22}$ sec. Maglic *et al* (1961) observed mass of 187 MeV, resonance width <25 MeV and life time > 4×10^{-23} sec and the distance travelled before its decay into three pions > 13 fermis.

The ω-meson has also been observed in a liquid deuterium bubble chamber exposed to a high energy beam of π^+-mesons.

$$\pi^+ + d \rightarrow p + p + \pi^+ + \pi^- + \pi^o \text{ or } p + p + \omega^o.$$

This experiment is accompanied by less background events than in proton-antiproton annihilation.

ρ-meson

Another quasi-particle is the rho (ρ)-meson. It is a two pions resonance particle. It can exist in three charge states ρ^+, ρ and ρ^o. These are produced in a fairly large fraction of high energy antiproton-proton annihilations.

$$\bar{p} + p \rightarrow \rho + \pi + \pi.$$

The most frequent decay mode of the charged p-meson is

$$\bar{p} \rightarrow p^+ + \pi^o.$$

The most probable decay of ρ^o is

$$\rho^o \rightarrow \pi^+ + \pi^-.$$

Other decays have small branching ratios.

$$\rho^o \rightarrow \pi^+ + \pi^- + \pi^o$$

$$\rho^o \rightarrow \pi^+ + \pi^- + \pi^o + \pi^o.$$

The observed rate of a dipion final state as a function of the total energy of the colliding beams (e^+ and e^- beams) shows a maxima at about 750 MeV. By calculating momenta of $\pi^+\pi^-$ pair by track chamber methods, the mass of ρ-meson was caiculated as 765 ± 4 MeV.

The width of ρ^o-meson is 100 MeV so that its life in sec it $\approx 10^{-33}$ sec.

As $\pi^+\pi^-$ states do not occurs hence T ≠ 2. The neutral and singly charged are observed, hence isospin T = 1. The assignment of I = 1 is made by analyzing the angular distribution of the decay products of ρ-particles emitted parallel or anti-parallel to the direction of the incident pion. It follows negative parity, considering that two decay pions have zero intrinsic spin and even intrinsic parity.

The resonance state of η particle, named as η' or x°, has mass 957.1 ± 0.6 MeV, spin 0, parity odd, isospin 0, hypercharge 0. The full width at half maximum is less than 4 MeV. Its decay modes arc

$$\eta' \rightarrow \eta + \pi + \pi \ (68\%)$$

$$\rightarrow \pi^+ + \pi^- + \gamma \ (32\%).$$

ELEMENTARY PARTICLE SYMMETRIES

Mendelyeev was able to predict correctly the atomic weights and other properties of then unknown elements by means of his periodic system. Similarly, the classification of the elementary particles has also met with success in predicting new particles. The method developed for the arrangement is based on a branch of advanced mathematics, known as *group theory*. It has been found that, in general, *the conservation law represents an invariance which corresponds to an appropriate symmetry operation.* The set of operators that represents the symmetry constitutes the group from which the theory gets its name. The irreducible representations of a group consist of a number of states, quantities or objects to which the symmetry operations are applicable. Thus the appropriate group operation can transform any one of these states into another in the same representation. The fundamental representation is the one containing the smallest number of states for the particular group.

If the system is invariant with respect to displacement in space linear momentum is conserved. Angular momentum is conserved if invariance is w.r.t. angular displacement and energy is conserved if it is w.r.t. time.

The simplest unitary group U (1) contains transformations which add a phase factor only to particle wave functions. The invariance under such transformations gives conservations of charge Q, baryon B, lepton L and hypercharge Y.

Unitary Symmetry [SU (2) Symmetry]

We know that proton and neutron are identical as far as the nuclear force is concerned, but differ in their electromagnetic interactions. Thus, it is possible to imagine a group of symmetry operators which could transform a neutron into a proton (or proton into a neutron) in the absence of an electromagnetic field.

The proton and neutron would then form the fundamental representations of the group. The existence of such a symmetry implies that something remains constant under the strong interaction. This is known as *isospin* and is 1/2 for proton as well as for neutron. The component of the isospin, T_3 is + 1/2 for the proton and – 1/2 for the neutron. The operators of the symmetry group thus change the co-ordinates of isospin in such a way as to reverse the sign of T_3 It can also be expressed as : *the strong interactions are assumed to be invariant under rotations in the isotopic spin space.*

The particular symmetry group applicable to isospin conservation is a form of *unitary symmetry known as a* U (2), which can be expressed by a set of 2 × 2 matrices. This group may be reduced to a *special unitary group S U* (2), which is also written as SU_2. It is special because a restriction reduces by unity the number of operators in the group. *The two dimensions* refer to the two basic states which make up the fundamental representatation in this case. The restriction of special reduces the number of operators 2 × 2 = 4 to three. The group is then said to have three generators.

By the use of the algebra of the SU (2) group it can be shown that all irreducible representations of the symmetry group consist of a multiplet of 2T + 1 states. All the members of the multiplet have the same isospin T and are essentially identical except for charge. If the symmetry was exact, i.e. isospin is strictly conserved, the components of a multiplet would differ in charge and T_3. The *SU (2) symmetry is violated by the electromagnetic interaction for which conservation of isospin is not applicable.*

The nucleon states |p >, | n > have anti-nucleon states |p >, |n >. Omitting | > brackets for clarity and separating the trace from traceless part, the combination of nucleon with an anti-nucleon may be represented as

$$\begin{pmatrix} p \\ n \end{pmatrix} \times (\bar{p}\ \bar{n}) \rightarrow \frac{\bar{p}p + n\bar{n}}{\sqrt{2}} \begin{pmatrix} 1 & 0 \\ 0 & 1 \end{pmatrix} + \begin{pmatrix} (p\bar{p} - n\bar{n})\frac{1}{2} & (p\bar{n}) \\ n\bar{p} & (p\bar{p} - n\bar{n})\frac{1}{2} \end{pmatrix}$$

The first term of r.h.s. represents singlet (η meson, T = 0, $I=0^-$) and the second term represents the triplet array (pions T = 1, I = 0^-). The second term can be written as

$$\begin{pmatrix} \frac{\pi^{o}}{\sqrt{2}} & \pi^{+} \\ \pi^{-} & \frac{\pi^{o}}{\sqrt{2}} \end{pmatrix}$$

Eightfold Way [SU (3) Symmetry]

The SU(3) theory is a generalization of the theory of isospin. This stands for *special unitary group in three dimensions*. The term, *three dimensions* refers to the three basic states which make up the fundamental representation in this case. In a three dimension unitary group there are, in general 3 × 3 = 9 operators, but the restriction of "special" reduces the number in eight. The group is then said to have *eight generators*. Gell-Mann has referred to the resulting group of symmetry operators as the *eightfold way, named for Buddha's Eightfold Path to Nirvana, comprising eight right actions*. Three of the generators apply to *three components of isospin, as in SU (2)* and a *fourth is associated with hypercharge. The remaining four also involve hypercharge in a different way.*

Application of the group algebra showed that the SU (3) symmetry should give rise to six supermultiplets, containing 1, 8, 8, 10, $\overline{10}$ and 27 members. The $\overline{10}$ multiplet is equivalent to the 10 but with hypercharges of opposite signs. In each of these supermultiplets the parity and intrinsic spin of members are the same, while the hypercharge and the isotopic spin are not same. Among above mentioned groups, 8 and 10 member groups are of particular interest.

In the case for B = 0 we may form particle anti-particle states to fill a 3 × 3 array :

$$\begin{pmatrix} p \\ n \\ \Lambda \end{pmatrix} \times \left(\bar{p}\,\bar{n}\,\bar{\Lambda}\right) \rightarrow \left\{ \begin{matrix} \frac{1}{3}\left(2p\bar{p} - n\bar{n} - \Lambda\bar{\Lambda}\right) & p\bar{n} & p\bar{\Lambda} \\ n\bar{p} & \frac{1}{3}\left(p\bar{p} + 2n\bar{p} + \Lambda\bar{\Lambda}\right) & n\bar{\Lambda} \\ \Lambda\bar{p} & \Lambda\bar{n} & \frac{1}{3}\left(-p\bar{p} - n\bar{n} + 2\Lambda\bar{\Lambda}\right) \end{matrix} \right\}$$

It can be identified with known spin zero mesons

$$\left\{ \begin{matrix} \left(\frac{\pi^o}{\sqrt{2}}\right) + \left(\frac{\eta}{\sqrt{6}}\right) & \pi^+ & K^+ \\ \pi^- & \left(\frac{\pi^o}{\sqrt{2}}\right) + \left(\frac{\eta}{\sqrt{6}}\right) & K^o \\ K^- & \bar{K}^o & -\frac{2\eta}{\sqrt{6}} \end{matrix} \right\}$$

The neutral η meson is now written as

$$\eta = \frac{\left(p\bar{p} + n\bar{n} + 2\Lambda\bar{\Lambda}\right)}{\sqrt{6}}.$$

There is in addition the symmetrical neutral combination or singlet

$$\eta' = \frac{p\bar{p} + nn + \Lambda\bar{\Lambda}}{\sqrt{3}}$$

Since these mesons are formed from fermion particle-antiparticle pairs, hence have odd parity. These eight particles with B = 0, and $I^p = 0^-$ should be arranged as :

one triplet with	Y = 0,	T = 1
one doublet with	Y = – 1,	T = 1/2
one doublet with	Y = –1,	T = 1/2 8 members
one singlet with	Y = 0,	T = 0

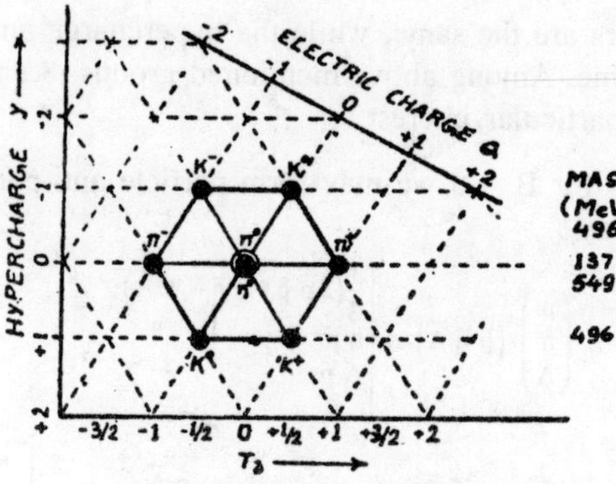

Fig. 3 : Boson Octet ($T^p = O^-$)

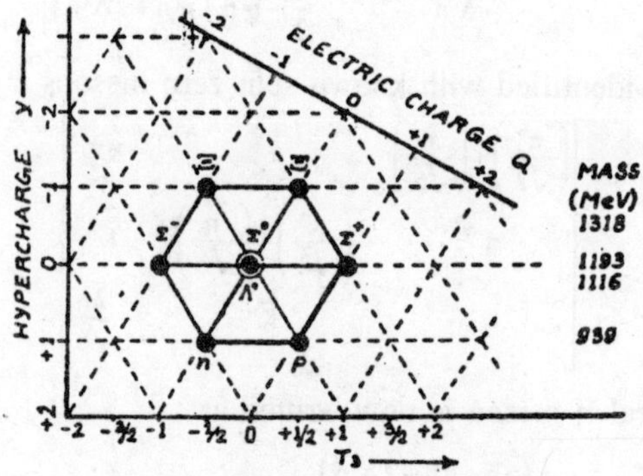

Fig. 4 : Baryon Octet ($J^P = 1/2^+$).

When this was first postulated there were seven ground state mesons. The missing meson was expected to have Q = 0, Y = 0, T = 0 and was predicted both by Gell-Mann and by Ne'eman. This eighth meson was named η-meson. Octects of metastable mesons (zero intrinsic spin) is shown in Fig. 3. In 1962, Gell-Mann and Ne'eman pointed out that the baryon also formed an octet array. For B = 1, the octet representation of SU (3) gives an octet array

$$\begin{Bmatrix} \left(\dfrac{\Sigma^o}{\sqrt{2}}\right)+\left(\dfrac{\Lambda}{\sqrt{6}}\right) & \Sigma^+ & p \\ \Sigma^- & \left(\dfrac{-\Sigma^o}{\sqrt{2}}\right)+\left(\dfrac{\Lambda}{\sqrt{6}}\right) & n \\ \Xi^- & \Xi^o & -\dfrac{2\Lambda}{\sqrt{6}} \end{Bmatrix}$$

These are eight in number (p, n, Λ, Σ^+, Σ^-, Σ^o Ξ^-, Ξ^o). For these $I^p = \frac{1}{2}^+$, B = 1, Y = 1 and T = 1/2 or 0. The eightfold way arrange- ment of this baryon octet is shown in Fig. 4.

For the octet mesons, Gell-Mann suggested a relationship among the average masses of the components, $\eta(M_0)$, π (M_1) and K $(M_{1/2})$, as

$$3M_0^2 + M_1^2 = 4M_{1/2}^2 \qquad ...(1)$$

Another interesting case is that of the boson octet with $I^p = 1^-$. The particles are :

$$K^{*o}\left(T_3 = -\frac{1}{2}, Y = 1\right), K^{*+}\left(T_3 = \frac{1}{2}, Y = 1\right), \rho^-(T_3 = -1, Y = 0),$$

$$\phi^o \,\&\, \rho^o\ T_3 = 0, Y = 0), K^{*-}\left(T_3 = -\frac{1}{3}, Y = -1\right),$$

$$\overline{K}^{*o}\left(T_3 = \frac{1}{2}, Y = -1\right).$$

For $I^P = 1^-$, a boson singlet ω^o (T = 0, Y = 0) has also been suggested. Neither the mass of ϕ^o nor ω^o fits with the expected value of M_0 in above, relation. The masses of ω and ϕ^0 are found approximately equally distant from the expected value (M_0 = 930 Me of mass. Gell-Mann and later Sakurai explained this by assuming ω^0 and ϕ^0 approximately the following mixtures of the octet and singlet states

$$\text{Real } \omega = \sqrt{\left(\frac{1}{3}\right)}\phi - \sqrt{\left(\frac{2}{3}\right)}\omega$$

$$\text{Real } \omega = \sqrt{\left(\frac{2}{3}\right)}\phi + \sqrt{\left(\frac{1}{3}\right)}\omega. \qquad ...(2)$$

Without really knowing which particle belongs in the octet state and which in the singlet, the ϕ^o has been arbitrarily placed in the octet.

Gell Mann derived a relationship among the average masses of the components of the four baryon multiplets in the intrinsic spin half octet, given as

$$2(M_N + M_\Xi) = 3\ M_\Lambda + M_\Xi, \qquad ...(3)$$

where M_N is the average nucleon mass and the average masses of the hyperons are indicated by the respective subscripts.

The relations (2) and (3) are special cases of the general formula, given by Okubo,

$$M\ (T, Y) = M_0\ [1 + aY + b\ \{T(T + 1) - 1/4Y^2\}], \qquad ...(4)$$

where a and b are constants for a particular multiplet.

The baryon octet for intrinsic spin 3/2 consists of one doblet ($Y = +1$; resonance states N^{*0} with $T_3 = -1/2$ and N^{**} with $T_3=1/2$); one doublet ($Y = -1$; resonance states Ξ^{*-} with $T_3 = -1/2$; and Ξ^{*0} with $T_3 = 1/2$); one triplet ($Y = 0$; resonance states Y^{*-} with $T_3 = -1$, y^{*0} with $T_3 = 0$ and Y^{*+} with $T_3 = +1$.) and one singlet ($Y = 0$, $T_3 = 0$ and member is $Y^{*(0)}$).

A more remarkable prediction was the case of the ten-fold baryon group with intrinsic spin 3/2. Theory predicted that there should be a group of ten members : $Y = 1$, $T = 3/2$, quartet (nucleon resonance denoted by N md exists in four charge states), $Y = 0$, $T = 1$, triplet (hyperon resonance Y_1^* which is containing excited Σ particles), $Y = -1$, $T = 1/2$, doublet (hyperon resonance which is equivalent to excited cascade particles Ξ), $Y = -2$, $T = 0$, singlet (a particle with charge -1).

In 1962, when the proposal was made to incorporate the N, Y_1, and Ξ resonances in a decuplet, the tenth particle was unknown. The existence of such a particle was predicted by Gell Mann and was named omega (Ω). The baryon decuplet for intrinsic spin 3/2 is shown in Fig. 5.

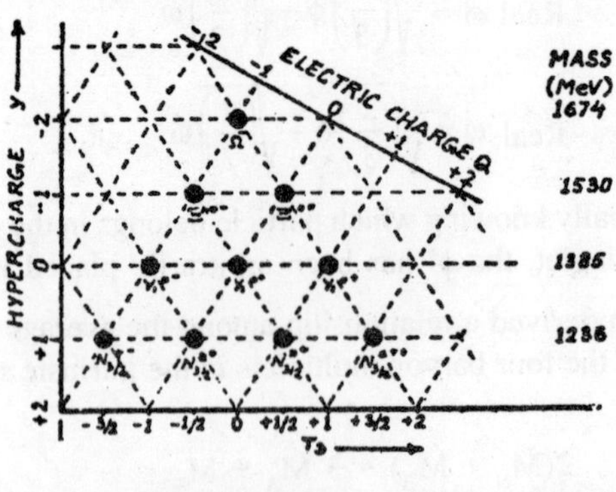

Fig. 5 : Baryon decuplet for intrinsic spin $3/2^+$.

Before the Ω^- had been discovered, the values of a and b (eqn. 51) could be found in order to fit the $Y = 1$, 0 and -1. Thus the mass of the Y = – 2 member was predicted as 1676 MeV, while the measured value is 1675 MeV. Although all the other members of the baryon decuplet for intrinsic spin 3/2 are resonant states, decaying by the strong interaction, Gell-Mann noted that the Ω^- particle should decay by the weak interaction. Possible decay modes are

$$\begin{aligned}\Omega^- &\rightarrow \Xi^- + \pi^o \\ &\rightarrow \Xi^o + \pi^- \\ &\rightarrow K^- + \Lambda^o\end{aligned}$$

It is dear that in these modes baryon number is conserved while isospin, hypercharge and strangeness are not conserved. These conservations only hold for strong interactions, hence the decay of Ω^- is by weak interaction.

In a recent development attempts have been made to combine ordinary spin with T and Y. The new group is described as SU (6).

QUARK THEORY

It is obviously desirable to find whether or not the multiplicity of particles can be built up from other similar units. Analysis has shown the possible existence of three basic units which could be the really *fundamental particles*. Zweing called the three particles *aces*, but the name *quark*, proposed by Gell-Mann, has become widely accepted. The elementary particles can be conceived (as far as isospin and hypercharge are concerned) as being built out of combinations of quarks. The quarks (a, b, c) are the basic states, which are represented as the basic three component column matrices.

$$a = \begin{pmatrix}1\\0\\0\end{pmatrix}, b = \begin{pmatrix}0\\1\\0\end{pmatrix}, c = \begin{pmatrix}0\\0\\1\end{pmatrix}.$$

The quarks a and b form a isospin doublet with T = 1/2 and T_3 = 1/2 and – 1/2, whereas the c quark is an isosinglet with $T_3 = 0$. The quarks can be interchanged with the unitary matrix, as

$$q_i = \sum_{j=1}^{3} U_{ij}\, q_j .$$

The unitary matrices U can be expressed in terms of Hermitian generators F_j as

$$U = \exp\left[i \sum_{j=1}^{8} \alpha_j F_j \right]$$

where α_1, α_2,...α_8 are parameters, and F_1, F_2... are F_8 eight independent traceless hermitian 3 × 3 matrices. The quantum numbers of the quarks, are given in the table 1.

Table 1

Quark	Q	T	T_3	B	Y	S
a	+2/3	1/2	1/2	1/3	1/3	0
b	–1/3	1/2	–1/2	1/3	1/3	0
c	–1/3	0	0	1/3	–2/3	–1

For the corresponding antiquarks, the numerical values are the same but the signs of Q, B, T_3. Y and S, are changed. The antiquarks are represented as row matrices

$$\bar{a} = (1, 0, 0), \quad \bar{b} = (0, 1, 0), \quad \bar{c} = (0, 0, 1),$$

Which transform as

$$\overline{q_i} = \Sigma \overline{q_i} (K)_{ij}$$

In associating particles with SU (3) representation, the baryons with integer baryon number must be associated with states formed from three quarks. Thus for baryons a a × b × c indicates 27 states (a singlet, two octets and a decuplet).

The mesons with zero baryon number must be form from one quark and one antiquark. Thus for mesons a × $\bar{b}$ gives 9 states (a singlet and an octet). Examples of meson formation are as follows :

$$\pi^+(a\bar{b}), \pi^o(\bar{a}a), \pi^-(\bar{a}b), K^+(a\bar{c}), K^o(b\bar{c}), K^-(\bar{a}c),$$

The baryons are each constructed from three quarks, no antiquarks. This permits both octet and decuplet representations, as observed actually. All the possible combinations of three quarks for baryons only are set out in the Table 2.

The quarks and antiquarks are assumed to be the interacting particles of spin 1/2 carrying the quantum numbers shown in table 4.5. The fact that the quarks have not been detected in nuclear collisions, either in cosmic radiation or in high energy laboratories, suggests that the quark must have very high mass.

Table 2

Quarks	Q	T	M	Y	B	S	Particle
a a a	2	1/2	2	1	1	0	–
		3/2	4	1	1	0	Δ^{++}
b b b	–1	1/2	2	1	1	0	–
		3/2	4	1	1	0	Δ^{-}
a a b	1	1/2	2	1	1	0	$p^{+}(N^{+})$
		3/2	4	1	1	0	Δ^{+}
a b b	0	1/2	2	1	1	0	$n^{o}(N^{o})$
		3/2	4	1	1	0	Δ^{o}
a a c	1	0	1	0	1	–1	–
		1	3	0	1	–1	Σ^{+}
a b c	0	0	1	0	1	–1	Λ^{o}
		1	3	0	1	–1	Σ^{o}
b b c	–1	0	1	0	1	–1	–
		1	3	0	1	–1	Σ^{-}
a c c	0	1/2	2	–1	1	–2	Ξ^{o}
b c c	–1	1/2	2	–1	1	–2	Ξ^{-}
c c c	–1	0	1	–2	1	–3	Ω^{-}

The quarks have no independent existence outside the hadrons like the phonons of solid stale physics. If the mesons arc composed of one quark interacting with one antiquark through a scalar potential, the total spin must be either 1 or 0. For the angular momentum L, the total parity P of the meson must be $(-1)^{L+1}$. The baryon states formed from three quarks must have total spin 3/2 or 1/2.

No evidence had been obtained that the quarks can exist. Attempts are now being made to find quarks in the fields of cosmic ray and large accelerator research but the difficulties are great. It has been suggested that meteorities might have collected quarks during their lifetimes in the planetary system of 4.5×10^{9} years. Such bits of cosmic material which have arrived on earth would perhaps contain charges less than the electronic charge. Attempts to detect these sub-electronic charges are now being tried but have get to be found.

ELECTROMAGNETIC STRUCTURE OF NUCLEONS

According to meson theory, the proton as well as the neutron, should be surrounded by a meson cloud and consists of a distribution charge of finite extent. *Electrons are used to probe the electromagnetic structure of the nucleon just as α-particles used to probe atomic structure.* Owing to the small size of the proton, electrons of small wave length (thus of high energy) are required for the experimental investigation of the charge distribution in the nucleons. After the completion of the linear accelerator at Stanford University, Hofstadter and others were able to determine nuclear radius and hence to study the structure of nucleons.

The first observations of a finite size of the proton were made by Hofstadter and McAllister (1955) by examining the scattering at 100 MeV electrons and found a very small effect of finite size. The experiments were further carried out at an electron energy of 236 MeV and found an rms radius of the proton to be equal to 0.78 ± 0.20 fermi. The numerical results were arrived at by assuming that

$$F_1 = F_2 \ (\equiv F_P)$$

as functions of q and effectively using the shape independent approximation

$$F = 1 - \left(\frac{q^2\alpha^2}{6}\right) + \ldots,$$

where 'α' is the rms radius of the charge or magnetic moment distribution. It is related more generally to a density distribution through the Fourier transform

$$F = \frac{4\pi}{q}\int_0^\infty \rho(r)\exp(i\vec{q}.\vec{r})\,d\tau,$$

which applies in the non-relativistic limit in which ρ (r) is the static density distribution and is a function of radius. At large angles, where tan θ/2 ≥ 1 and at high energies, it becomes possible to distinguish moment scattering. Hence it becomes possible to find F_1 and F_2 separately. Chambers and Hofstadter (1956) carried out a detailed series of experiments on the proton extending to an energy of 550 MeV with scattering angles lying between 30° and 135° in the lab. frame. They found a very large deviations from point scattering. They ratio F_1/F_2 may be found independently of any model by comparing two cross sections at two different energies and different angles, such that the momentum transfer q is the same at both cross section determinations. $F_1/F_2 = 1.1 \pm 0.2$ and $q^2 = 4 \times 10^{30} m^{-2}$ found from the above experiments.

Using the fact that magnetic moment scattering falls off more slowly at large angles than charge scattering, high energy data may be analyzed to find $F_1 + kF_2$. Since $k = 1.79$, the data can yield F_2, if F_1 is known even approximately. At small scattering, magnetic scattering is essentially zero and hence F_1 may be investigated alone. This procedure shows that for an exponential model, the charge r.m.s. radius must be close to 0.8 fermi when the magnetic r.m.s. radius is allowed to vary from 0.60 fermi to 1.0 fermi.

From the scattering at 900 MeV and at 1.3 BeV, we find that for high four momentum transfers the charge and anomalous moment from factors differ from one another in the sense that corresponds to a wider distribution for the magnetic moment than for the charge. The equations of theoretical curves show that the proton root mean radii are given as

$$r = r_{pc} = 0.88 \text{ fermi and } r_2 = r_{n\mu} = 0.95 \text{ fermi.}$$

Since neutron targets are not available, the information on the form factors of the neutron is obtained from the scattering of fast electrons by deuterons. Data upto 1.3 BeV have been obtained. The root mean square radii obtained from these results are :

$$r_1 = r_{nc} = 0.00 \text{ fermi and } f_2 = r_{n\mu} = 0.87 \text{ fermi.}$$

Charge distributions in the proton and in neutron obtained from above calculations are shown in Fig. 6.

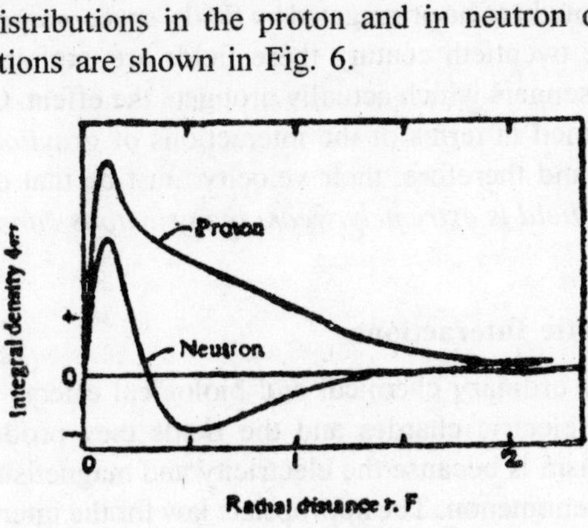

Fig. 6 : The proton and neutron charge density distribution.

The proton has a finite extent as if there were a finite probability of finding it decomposed into a neutron and positive meson cloud. The structure of the neutron can be interpreted as due to the decomposition into a proton surrounded by a negative muon cloud, but apparently produces a positive charge density at larger distances.

PARTICLE INTERACTIONS

The interactions among elementary particles can be classified into following four types :

The Gravitational Interaction

The first force that any of us discover is gravity. It holds the moon and earth together, keeping the planets in their solar orbits and binds stars to form our galaxy. Newton gave a formula $F = Gm_1m_2/r^2$ for the interaction between two masses. The gravitational effect does not depend on the colour, size, charge, velocities, spin and angular orientation but depends on the magnitude of the inertia. The gravitational force between two nucleons separated by a nucleon diameter is

$$F = G\frac{m_1 m_2}{r^2} = 6.7\times10^{-11}\frac{\left(1.7\times10^{-27}\right)^2}{\left(10^{-15}\right)^2} \approx 2\times10^{-34} \text{ newtons} \quad ...(1)$$

and the gravitational attraction is only about 2×10^{-49} joule. Hence we sec that it plays no role in particle reactions. In the ninteenth century the forces were thought to be propagated by fields, space warped for particular effects. In the twentieth century these fields are explained in terms of agents or messengers which actually propgate the effect. Gravitation can thus be explained in terms of the interactions of *gravitons*. Their mass must be zero and therefore, their velocity must be that of light. *As the gravitational, field is extremely weak, the gravitons cannot be detected in laboratory.*

Electromagnetic Interactions

All of the ordinary chemical and biological effects are due to the interaction of electric charges and the fields they produce. The term electromagnetism is because the electricity and magnetism are both part of the same phenomenon. The appropriate law for the interaction of point charges bears the name of Coulomb ($F = q_1q_2/4\pi\epsilon_0 r^2$). For two protons, 10^{-15} metre apart, the repulsion force will be $9 \times 10^9 \times (1.6 \times 10^{-19})^2/(10^{-25})^2 \approx 30$ newtons. It is about 10^{35} times greater than the gravitational attraction caused by the mass. The energy released by the complete separation of these protons would be 3×10^{-14} joules.

If the particles are not at rest but are moving, the field will not only be an electric field but would be new one depending on the velocity and magnitude of the charge. When the charge is accelerated, the energy is radiated out in the form of an electric and megnetic pulses. This energy comes from the agent which accelerates the charge. The pulse is called a *photon* and travels with the velocity of light. If the source charge is accelerating in an oscillating fashion, the propagated signal will consist of successive waves of electric and magnetic fields or the *radio-photons*. Thus we see that the photons are emitted and reabsorbed by a charge. *Interaction between two charged particles consists of an exchange of these photons*. The strength of the electromagnetic interact n is given by the dimensionless fine structure constant $\alpha(= e^2/\epsilon_0 \hbar_c = 1/137)$, and is due to photon exchanges.

The electromagnetic interaction is charge dependent. In terms of isobaric spin, the interaction depends on T_s and is governed by the isospin rule $\Delta T = 0, \pm 1$. All Other quantities such as charge, baryon number, lepton number, hypercharge, parity, strangeness number are conserved.

The capture of photon can effect the production of mesons or hyperons by an electromagnetic interaction

$$\gamma + p \qquad \begin{matrix} \pi^o + p \\ \\ \Lambda^o + K^+. \end{matrix}$$

An example of a irradiative capture reaction is

$$\pi^- + p \rightarrow n + \gamma.$$

The neutral particles such as

$$\pi^o \rightarrow \gamma + \gamma, \ \Sigma^o \rightarrow \Lambda^o + \gamma,$$

$$\eta^o \rightarrow \pi^+ + \pi^- + \pi^o \qquad \eta^o \rightarrow \gamma + \gamma,$$

decay electromagnetically since these processes involve no change of strangeness The decay processes such as $\Sigma^+ \rightarrow p^+ + \gamma$ are forbidden because the change $\Delta S = 1$ is required. The paradox that the decay of netural particle is by electromagnetic nteractior is resolved by introducing as an intermediate step in the overall reaction. The virtual production of a *nucleon-anti-nucleon (or electron-positron)* pair. Thus we have

$$\pi^{o} \xrightarrow{\text{strong}} \underset{\text{virtual}}{(N+\overline{N})} \xrightarrow{\text{electromagnetic}} \gamma + \gamma.$$

The process of mutual annihilation of particles and anti-particles is an example of electromagnetic interaction

(c) Strong Interaction

The strong nuclear interaction is independent of the electric charge. The force is same between p-p and n-n. For this purpose the proton and neutron are, one but in different electric charge states. Strong interactions involve mesons and baryons. The range is very much shorter than that of gravitational or electromagnetic interaction. Strong interaction energy falls off rapidly when the distance between two particles increases. Yukawa in 1935 predicted the existence of heavy quanta, which played the same role in nuclear forces (or strong forces) as photons in electromagnetic ones. From estimates of the range of nuclear forces. Yukawa predicted that the new particles, called mesons, should have a mass of the order of 200 to 300 electron masses. In the chapter of nuclear forces, we outlined an elementary theory of pion-nucleon interaction and introduced the concept of a nucleon charge g analogous to the electric charge e. The strength of the nuclear interaction is represented by the magnitude of the dimensional coupling constant $\gamma^2/4\pi\hbar c$ ($\approx$ 14). It is about a thousand times the electromagnetic coupling constant α.

Strong interactions between elementary particles are responsible for the total cross sections as a function of energy. The strong interaction is a short range force ($\approx 10^{-15}$ m), conserves baryon number B, charge Q, hypercharge Y, parity π, isospin T and its component T_s. It is responsible for kaon production, however the decay of mesons, nucleons and hyperons proceeds by an electromagnetic or weak interactions.

(d) Weak Interaction

The weak interaction is responsible for the decay of strange and non-strange particles and for non-leptonic decay of strange particles. The numerical constant, which is characterstic of the weak interactions, is obtained from Fermi's theory of decay. Its value is $g_F = 1.41 \times 10^{-62}$ Jm^2. In analogy with the expression for the other interactions, the dimensionless weak interaction coupling constant is of magnitude

$$\left[\frac{gF^2}{(\hbar_c)^2}\right]\left[\frac{m_{rc}}{\hbar}\right]^4 \simeq 5 \times 10^{-24}.$$

Consider the reactions which do not involve a change of strangeness and yet which must be due to weak interaction. The neutron decay is the proto-type of all the β-decays :

$$n \rightarrow p + e^- + \overline{\nu_e}.$$

The nature of such an eqn. is that the reaction can go in either direction so long as energy is conserved find that any participant can be replaced on opposite side by its anti-particle, i.e. $p \rightarrow n + e^+ + \nu_e$. Another variation of this four fermion interaction is the proton capture of an anti-neutrino $\left(\overline{\nu_e} + p \rightarrow \nu + e^+\right)$. Another example of the four-fermion interaction is the muon decay $\left(\mu \rightarrow e^- + \nu_\mu + \overline{\nu_e}\right)$. There is also a coupling among muons, nucleons and neutrinos, thus

$$\mu^- + p \rightarrow n + \nu_\mu.$$

There are however other types of weak interactions which require coupling between other pairs of fermions.

$$\Lambda^o\ (S = -1) - \pi^-\ (S = 0) + p\ (S = 0).$$

There is no neutrino involved here, strangeness changes by + 1 and only two fermions are involved. For four fermions one can assume this decay as through a virtual stage $\left(\Lambda^o \rightarrow \bar{p} + n^o + p^+\right)$. In the first vitual step, four fermions are involved. In the next stage, the strong nuclear forces come into play. There are restrictions whether one transition is forbidden or allowed : *In a weak interaction involving change in strangeness of baryons or mesons, the change in strangeness must be equal to the change in charge.* The change in isospin T and its component T_z may be nonzero. The strangeness and isospin are not meaningful for leptons, and are useful when hadrons are involved in the weak interactions. The lepton number is conserved in these interactions. The parity is not conserved, but CP and CPT are conserved.

Similar to the graviton for gravity, photon for elecromagnetism and mesons for the strong nuclear force there is one agent for the weak interaction. It would also be a boson of mass above 800 MeV and is named as the *intermediate charged vector Boson* and given the symbol W. Its half life against decay into electron neutrino or muon-neutrino would be less than 10^{-17} sec.

One associates neutrino and anti-neutrino exclusively with the weak interaction, just as one associates photons with em-interaction. One particle may respond to the different types of interaction' e.g., pion scattering is effected by a strong force, pion radioactive capture by an electromagnetic interaction and decay by a weak interaction.

We can thus write these interactions as :

Table 3 : Comparison of the four basic interactions

Field	Relative magnitude	Associated particles	Character istic time
Strong interaction	1	Pion, kaon	10^{-23} sec.
Electromagnetic	10^{-3}	Photon	10^{-20} sec.
Weak interaction	10^{-13}	Intermediate boson	10^{-10} sec.
Gravitational	10^{-39}	Graviton	10^{-16} sec.

CONSERVATION LAWS

The behaviour of the elementary particles is restricted by a number of conservation laws or invariance principles. That is to say, certain properties of representative physical quantities must remain unchanged in any process. The most familiar quantities in large scale experiments that are conserved in all (strong, electromagnetic and weak) interactions are:

1. Conservation of Angular Momentum

The conservation of angular momentum includes both types (orbital and spin) of angular momentum together. The first is given by the motion of the object as a whole about any chosen external axis of rotation. The second is the intrinsic angular momentum of each object about an axis through its own centre of mass. Strongly interacting fermions have a half integer spin (s = 1/2 for Ξ, Σ, Λ, n and p; s = 3/2 for Ω), strongly interacting bosons (η, K, π) have s = 0, weakly interacting fermions (leptons μ, e, ν_e, ν_μ) have s = 1/2, massless bosons (electromagnetic interacting γ-rays) have s = 1 and gravitions have s = 2.

2. Conservation of Energy

Conservation of energy on other hand seems more complicated with elementary particles because a large fraction of the total energy is oftenly interchanged between rest energy associated with mass and kinetic or potential energy. The sum of these three, the total energy is always conserved in any reaction. For example the decay reaction $K^o \neq \pi^+ + \pi^- + \pi^o + \pi^o$ is forbidden because the rest energy of the K^o is not great enough to make four pions even if they all could be make at rest. However, the reaction $K^o \rightarrow \pi^+ + \pi^- + \pi^o$ is allowed.

3. Conservation of Charge

The most familiar of the conservation laws is the conservation of electric charge. The charge is conserved in all processes and no exceptions are known. We note that all elementary charges are 0, or –1; multiple charges are not found.

4. Conservation of Baryon Number

The number of baryons minus the number of anti-baryons is conserved. In other words the net baryon number in any process always remains unchanged. All normal baryons such as p^+, n^o, Λ^o, Σ^+, Σ^-, Σ^-, Ξ^-, Ξ^o and Ω^- have a baryon number of +1, the corresponding anti-particles known as anti-baryons have a baryon number of – 1. All the mesons have a baryon number of zero. For example, the reaction $A^o \rightarrow \pi^+ + \pi^-$ is *allowed* because the baryon Λ^o is replaced by the baryon p^+, keeping the total number of baryons constant. The reaction $\Lambda^o \rightarrow + p^- + \pi^+$ is *forbidden* because one baryon is replaced by one anti-baryon changing the baryon number by – 2.

5. Conservation of Lepton Number

The number of leptons minus the number of anti-leptons is conserved. In other words *the net leptons number in any process always remains conserved.* The ordinary electron, negative muon and neutrinos all have a lepton number + 1, the corresponding anti-particles known as antileptons have a lepton number of –1. The reaction $\mu^+ \rightarrow e^+ + n + \bar{\nu}$ is allowed because a lepton number of –1 is replaced by a lepton number (–1) + (+1) + (–1) = – 1.

The reaction $n \rightarrow p + e^- + \bar{\nu}$ is allowed by both conservation of baryons and conservation of leptons. No exception to the rules of conservations of baryons and leptons has been found in the many elementary particle reactions so far studied one by one. Conservation of leptons has a significance for strong interactions. There are other conservation laws which are not applicable to weak interactions. The property that is conserved in strong interactions only is known as isospin. Other properties which are not conserved for all the three interactions, but are conserved in one or two interactions only, are hypercharge, *strangeness, purity*, invariance under charge conjugation and invariance under CP conjugation.

1. Conservation of Isospin

According to the ordinary idea of isotopic spin, each nuclear particle posseses a certain total isotopic spin T and each possible projection of this isotopic spin along a certain axis T_3, appears to us as a different charge state of the corresponding particle. In the case of nucleons, T = 1/2 and the 2T + 1 = 2, possible values of T_3 are + 1/2 (for the proton state) and – 1/2 (for the neutron state). For the pions, T = l and so there are 2T + 1 = 3 charge states. The triplet consists of π^+, π^0 and π^- particles and the values of T_3 are +1.0, –1, respectively.

B is +1 for the proton, neutron and hyperons and is zero for the pions. Inserting the value of T_3 and B, one obtains :

$$Q_p = +\frac{1}{2}+\frac{1}{2}=+1,\ Q_n = -\frac{1}{2}+\frac{1}{2}=0,$$

$$Q\pi^+ = +1+0 = 1,\ Q\pi^o = 0+0 = 0,\ Q\pi^- = -1+0 = -1.$$

Analytically, if B is the baryon number, T the isotopic spin quantum number, T_3 the component of T and Q the charge in units of the electron charge, the relation between these quantities is

$$Q = T_3 + \frac{B}{2}. \qquad ...(2)$$

Isospin numbers are associated with *hadrons* (particles that can exhibit strong interactions) but not with leptons. The isospin component T_3 is conserved in both strong and electromagnetic interactions but not in weak interaction.

2. Conservation of Hypercharge

A quantity called *hypercharge* (the twice the average charge of the members of the group), is also conserved in strong and electromagnetic interactions. For example, for the triplet π^+, π^0, π^-, average charge is zero and hence all these three mesons have a hypercharge of zero. The hypercharge of the pair of the particles K^+ hnd K^0 is +1 and that of the pair of anti-particles K^- and $\overline{K}^0$ is – 1. Thus the alternative definition is that it is twice the difference between the actual charge Q and the isospin component T_3 of a particle. Thus hypercharge

$$Y = 2(Q - T_3). \quad ...(3)$$

3. Conservation of Strangeness

The concept of strangeness has found wide application in particle physics. It is an additional quantum number which describes the interactions of elementary particles. It has been chosen in such a manner that it becomes zero for all the well known particles (non-strange particles). Rochester and Butler, in 1947, at the University of Manchester arranged a magnetic cloud chamber and after about a year of operation reported certain new types of forked or V shaped tracks in cloud chamber photographs of cosmic rays. These new V-particles were also investigated intensively by cloud chamber groups from various institutions for several years. By 1953, at an International Conference on Elementary particles at Bagneres a decision was taken to name those particles of mass greater than π-mesons, but smaller than protons, *K-particles*, while those particles whose masses were greater than protons were to be termed hyperons.

One of the most common V-particles (Λ^o) was neutral and decayed ($\Lambda^o \rightarrow p + \pi^-$) in time 2.5×10^{-10} seconds. The question arised if the Λ^o could interact strongly, why did not its decay go via strong interactions with a life time of $\sim 10^{-23}$ sec instead of the observed 10^{-10} sec? To explain this Pais suggested the *hypothesis of associated production. The V-par tides can interact strongly and, therefore, are produced only in pairs, once separated each number can decay into ordinary particles only through the weak interaction. A typical example is*

$$\pi^- + p^+ \rightarrow \Lambda^o + K^o$$

$$\rightarrow p^+ + \pi^- \rightarrow \pi^+ + \pi^-.$$

Both the associated creation of the strange particles and their individual stability against immediately decay were the features that earned them the title *strange*. Both features can be explained by insisting that the total *strangeness* must remain constant in fast particle reaction.

If the baryons are arranged in columns according to their electric charge (plus under plus, minus under minus, neutral under neutral). The electric charge centres do not thus occur in the same vertical line. The electric charge centre of the nucleon is at + 1/2, halfway between p and n. The charge centre of Λ^o is at 0. The triplet sigma is centred at 0, but the doublet xi is centred at – 1/2. The Ω^- singlet is at charge – i. If we take the charge centre of the nucleon doublet arbitrarily to be reference origin, then we have

For Λ^o ... $\Delta Q = Q_\Lambda - Q_N = -1/2$; for Σ-hyperons ... $\Delta Q = Q_\Sigma - Q_N = -1/2$: for Ξ-doubtlet...$\Delta Q = -1$ for Ω^- ... $\Delta Q = -3/2$.

By defining strangeness quantum number as

$$S = 2\Delta Q, \qquad ...(4)$$

We obtain S = 0 for the nucleons and non-zero for hyperons (S = – 1, for Λ^o and Σ's – 2 for Ξ's and – 3 for Ω^-). Once a hyperon or K-meson is produced and gets beyond the influence of the collision, it can decay. As the decay products have strangeness zero hence the strangeness remains conserved in fast nuclear progresses. Once produced and separated from each other, they must wait for some weaker interaction to allow them to decay. Strangeness conservation governs only the strong interactions and not the weak. Even the weak interactions have some respect for strangeness. It is found experimentally that weak decays change strangeness as little as possible.

$\Delta S = \pm 1$. No example with $\Delta S = \pm 2$ has been seen.

The examples are :

$$\Sigma^+ \rightarrow \Lambda^o + e^+ + \nu_e, \qquad \Delta S = 0$$

$$\Sigma^- \rightarrow n + e^- + \overline{\nu_e}, \qquad \Delta S = 1$$

$$\Lambda^o \rightarrow p + e^- + \overline{\nu_e}, \qquad \Delta S = 1.$$

M. Gell Mann in U.S.A. and T. Nakano and K. Nishijima in Japan independently suggested this strangeness quantum number. They suggested a scheme, known as Gell-Mann and Nishijima scheme after their names, for this quantum number. In this relation (2) is replaced by

$$Q = T_3 + \frac{1}{2} Y = T_3 + \frac{1}{2}(B + S), \qquad ...(5)$$

where $Y = S + B$ ...(6)

Since B is conserved always, the strangeness like the hypercharge is conserved in strong and electromagnetic interactions.

To cover a whole range of strongly interacting particles, the formalism has also been extended to mesons. Since the hyperons all have negative strangeness, they can be produced only in association with K-mesons of positive strangeness. On the assumption that strangeness is a conserved quantity in a reaction such as

$$p^+ + p^+ \rightarrow p^+ + \Lambda^o + K^+$$

S: $\quad 0 + 0 \rightarrow 0 + (-1) + S_k$,

strangeness $S_k = +1$. Similar argument gives strangeness S = 1 for neutral kaon. From the associated production reaction

$$\pi^- + p \rightarrow \Lambda^o + K^o$$

S: $\quad S_\pi + 0 = -1 + 1,$

it follows that $S_\pi = 0$. It also applies to π^+, π^o and η^o mesons. Since for mesons the baryon number B = 0, hence strangeness S = hypercharge Y.

Thus we see that strangeness is not an independent new quantity, but is related to a combination of Q, T_2 and B, each of which is regulated by conservation laws.

Let us now see what kinds of particle type can be formed by various choices of T, B and S. *If S = 0, there are three possibilities,*

(a) B = 0, T = 0 yields Q = 0 (natural meson), η^o meson.

(b) B = 0, T = 1 yields Q = + 1, 0, – 1 (pions).

(c) B = 1, T = 1/2 yields Q = + 1, 0 (nucleons).

If S = 1, the multiple charge can be avoided only when

(a) B = 1, T = 0 and Q = + 1 (Baryon singlet).

(b) B = 0, T = 1/2 and Q = +1, 0 (K^+, K^o).

If S = – S, the multiple charges can be avoided only when

(a) B = 0, T = 1/2 and Q = 0, – 1 ($\overline{K}^o$ and K^-).

(b) B = 1, T = 0 and Q = 0, (singlet baryon Λ^o).

(c) B = 1, T = 1 and Q = +1, 0, – 1 (Σ+, Σ^o and Σ^-).

If S = – 2, the multiple charges can be avoided only when

(a) B = 1, T = 1/2 and Q = 0, – 1 (hyperons Ξ^o, Ξ^-).

If S = – 3, the multiple charges can be avoided only when

(1) B = 1, T = 0 and Q = – 1 (Ω-hyperon).

All these results are summarised in the following table 4:

In the case of hadrons, the strangeness must be conserved ($\Delta S = 0$) for fast reactions. The decays of kaons and hyperons are very slow because they involve a breakdown of strangeness conservation.

Table 4

S	B	T	T_2	Q	Particle	
0	0	0	0	0	η^o	
	1	1/2	1/2	+1	p	
			−1/2	0	n	
	0	1	+1	+1	π^+	
			0	0	π^o	
			−1	−1	π^-	
1	1	0	0	1	Baryon (Not detected)	
	0	1/2	+1/2	1	K^+	
			−1/2	0	K^o	
−1	0	1/2	+1/2	0	K^-	
			− 1/2	−1	K^-	
	1	0	0	0	Λ^o	
	1	1	+1	+1	Σ^+	
			0	0	Σ^o	
			−1	−1	Σ^-	
	−2	1	1/2	1/2	0	Ξ^o
			− 1/2	−1	Ξ^-	
−3	1	0	0	−1	Ω^-	

The decays $\Xi^- \rightarrow n + \pi^-$ and $\Xi^o \rightarrow p + \pi^-$ which involve $\Delta S = 2$ would be expected to be exceptionally slow, and the decays $\Xi^- \rightarrow \Lambda^o$ π^- and $\Xi^o \rightarrow \wedge^o + \pi^o$ with $\Delta S = 1$ to be slow. Since charge and baryon number are always conserved hence for weak decay processes, eqn (5) gives

$$|\Delta S| = 2\,|\Delta\, T_2|$$

or $$|\Delta T_3| = \frac{1}{2} \qquad \text{...(7)}$$

As T_3 is the component of total isospin along a particular direction, hence the general form of eqn (7) for weak decays is

$$|\Delta T| = \frac{1}{2}. \qquad \text{...(8)}$$

Let us apply conservation laws to high energy pion-nucleon collisions which often give large quantities of kaons. Examples of possible equations are

$$\pi^+ + n^o \rightarrow \Lambda^o + K^+$$

$$\rightarrow K^o + K^+.$$

Using the conservation of baryons and strangeness, we see that the second reaction violates baryon conservation and, therefore, cannot occur. Similarly equation $\pi^- + p^+ \rightarrow \Lambda^o + K^o$ is possible whereas $\pi^- + p^+ \rightarrow \Lambda^o + \pi^o$ is not possible by strangeness violation.

Let us consider equation $p^- + p^+ \rightarrow 2\pi^+ 2\pi^- + \pi^o$. Applying various conservation laws and remembering that the pairs of pions are ejected with opposite isospins, we have

$$Q = -1 + 1 \rightarrow 2 - 2 + 0, \qquad \therefore \delta Q = 0$$

$$B = -1 + 1 \rightarrow 0 + 0 + 0, \qquad \therefore \delta B = 0$$

$$T = \frac{1}{2} + \frac{1}{2} \rightarrow 0 + 0 + 1, \qquad \therefore \delta T = 0$$

$$Y = -1 + \rightarrow 0 + 0 + 0, \qquad \therefore \delta Y = 0$$

$$S = 0 + 0 \rightarrow 0 + 0 + 0, \qquad \therefore \delta S = 0.$$

4. Charge Conjugation

Charge conjugation is defined as the interchange of particles and anti-particles. It does not simply mean a change over the opposite electric charge or magnetic moment, the sign of other charge quantum numbers [hypercharge Y, baryon number B, lepton numbers (l_e, l_μ)] is also reversed without changing mass M and spins. Thus a unitary operator, also known as charge conjugation operator C, satisfies the following relations :

$$CQC^{-1} = -Q,\ CYC^{-1} = Y,\ CBC^{-1} = -B,\ Cl_eC^{-1} = -l_e$$

and $Cl_\mu C^{-1} = l_\mu$.

Some elementary particles e.g., γ, n° – mesons and the positronium atom ($e^- + e^{+\prime}$) are transformed into themselves by charge conjugation. They are their own anti-particles. These are known as *selfconjugate or true neutral particles*. The neutron (B = 1, Y = 1) and K° -- mesons (Y = 1, B = 0) are not invariant under C.

A system is said to possess charge conjugation symmetry or to be invariant under charge conjugation if the system (or the process) is such that it is impossible to know that it has undergone charge conjugation. For example, the operation C converts the negative pion decay

$\left(\pi^- \rightarrow \mu^- + \overline{\nu_\mu}\right)$ into the positive pion decay $\left(\pi^+ \rightarrow \mu^+ + \nu_\mu\right)$, since the π^+ is the anti-particle of π^-. Until about two decades ago it was believed that the entire universe is invariant under C. In the end of 1956, experiments revealed that weak interactions violated it. It turned out that the μ^+ and μ^- decay electrons have angular distributions of opposite asymmetry : that the π^+ and π^- decay muoni have opposite polarizations and that while a free neutrino is left handed an anti-neutrino is right handed. Charge conjugation applied to a free moving neutrino then results in a process which does not exist in nature.

The charge conjugate of the Dirac equation, which corresponds to the wave functions of positrons, has the form

$$C = \text{i.e.}^{i\phi} \begin{pmatrix} 0 & \sigma_y \\ \sigma_y & 0 \end{pmatrix} = e^{i\phi}\, i\, \alpha_y. \qquad ...(9)$$

The phase ϕ is arbitrary. For zero phase

$$C = i\alpha_y, \qquad ...(10)$$

where α_y is a square matrix and σ_y is the Pauli spin matrix having value

$$\sigma_y = \begin{vmatrix} 0 & -i \\ i & 0 \end{vmatrix} \qquad ...(11)$$

For a single photon state

$$C\,|\gamma\rangle = -\,|\gamma\rangle. \qquad ...(12)$$

A state vector for n-photons

$$C\,|n\gamma\rangle = (-1)^n\,|n\gamma\rangle. \qquad ...(13)$$

Thus for n-quanta the eigenvalue of C is $(-1)^n$. Since π^o mesons decay through electro magnetic interaction into two photons $\pi^o \rightarrow 2\gamma$, it follows that the π^o-meson is in eignstate of C with eignvalue +1, *i.e.*

$$C\,\pi^o\rangle = |\pi^o\rangle$$

On the other hand $|\pi^o\rangle$ and $|\pi^-\rangle$ and not eignstates of C as

$$C\,|\pi^+\rangle = -\,|\pi^-\rangle \text{ and } C|\pi^-\rangle = -\,|\pi^+\rangle.$$

As the triplet spin state is symmetrical and the singlet spin state anti-symmetrical, hence to exchange an electron with a positron we must induce factor $(-1)^{s+1}$ as well as a factor $(-1)^l$. Thus we have

$$(-1)^{l+s+1}\, C = -1 \text{ or } C = (-1)^{l+s}.$$

This relation gives that the singlet ground state ($l = 0$) decay into two photons (C = 1) and the triplet ground state decays into three photons.

5. Space-Inversion Invariance (Parity)

The parity principle says that there is a symmetry between the world and its mirror image. This may be defined as reflection of every point in space through the origin of a co-ordinate system $x \to -x$, $y \to -y$ and $z \to -z$. If a system or process is such that its mirror image is *impossible io obtain in nature*, the system of procéss is said to violate the law of parity conservation.

Human body is a good example of mirror symmetry. The body of a car is symmetric except for the position of the steering wheel. If we were looking at the mirror image of a normal car, it seems to violate the symmetry but it is not the case as it is also possible to design a car with the steering wheel on the other side. The mirror view of a printed pure looks wrong. But there is nothing impossible about it. A printer could design inverted type and produced a page. One can read the page from right to left. The type of printing is not unnatural but is unconventional and unfamiliar.

All phenomena involving strong and electromagnetic interactions alone do conserve parity. In these cases the systems can be classified by the eigen values of the parity operator P. For a single particle Schrodinger wave function ψ, the result of the parity operation is

$$P\,|\psi(x)> = e^{i\alpha}\,|\psi(+x)\,\Delta, \qquad ...(14)$$

As α is an arbitrary real phase, hence can be set equal to zero.

$$P|\psi(x)> = |\psi(-x)> \qquad ...(15)$$

$$\text{and } P^2|\psi(x)> = |\psi(x)>. \qquad ...(16)$$

It shows eignvalues of P as + 1 or – 1.

The parity of the photon depends upon the mode of transition, it is due to the change of the sign of electromagnetic current j under the parity operation. The nucleons and electrons are assigned *positive* or *even* intrinsic parity. The pions have *negative or odd* parity as they involve in strong interactions with nucleons. K-mesons and η°-meson have negative parity. $\Lambda^\circ - \Xi^- -$, $\Sigma - \Omega$-hyperons have positive intrinsic parity. All anti-particles of spin 1/2 are of opposite parity to the corresponding particle, while the bosons and their anti-particles have the same parity.

The conservation of parity requires that the Hamiltonian of a free system commute with the parity operator.

$$(PH - HP) = 0. \qquad ...(17)$$

The transition probability must be scalar it may contain pseudoscalar operator (I.p). The conservation of parity prevents the mixing of even and odd operators in the amplitude. Thus for the non-conservation of pairty in β-decay, the transition probability must contain both scalar and pseudoscalar terms, or the number of electrons emitted parallel and anti-parallel to the spin of the source should be different.

The weak decay of the K-mesons, which was difficult to reconcile with parity conservation and known as the τ-θ puzzle, was explained by Lee and Yang. They suggested that *the weak interaction was not invariant to space reflection.* In 1956, Wu and others, using polarized Co^{60} nuclei, found that the direction of emission of electrons in the transformation to Ni^{60} was preferentially opposite to the spin direction. The value of the pseudoscalar I. p, where I is the nuclear spin and p the electron momentum, was measured and found to be different from 0.

6. Combined Inversion (CP)

Landu (1956) advanced a hypothesis to the effect that any physical interaction must be invariant under simultaneous reversal of position coordinates and change over from particles to anti-particles. For example a neutrino has a definite helicity and its parity conjugate has opposite helicity. The charge conjugate of the neutrino also has opposite helicity. *Thus under the combined operation* PC (*or* CP) *the neutrino changes to anti-neutrino.* The combined operation also known as *combined parity* (charge and space) *is conserved in most of the known physical processes.* Let us consider the decay of the positive pion,

$$\pi^+ \rightarrow \mu L^+ + \nu_\mu L.$$

Here subscript L indicates that neutrino and +ve muon fly apart with left handed spin. As the C-inversion changes particles into anti-particles and vice-versa, whereas the P-inversion converts left handed motion to right handed motion. Hence

C-inversion : $\pi^- \rightarrow \mu L^- + \nu_\mu L$ impossible process

P-inversion : $\pi^+ \rightarrow \mu R^+ + \nu_\mu R$ Impossible process

CP-inversion : $\pi^- \rightarrow \mu R^- + \nu_\mu R$ possible process

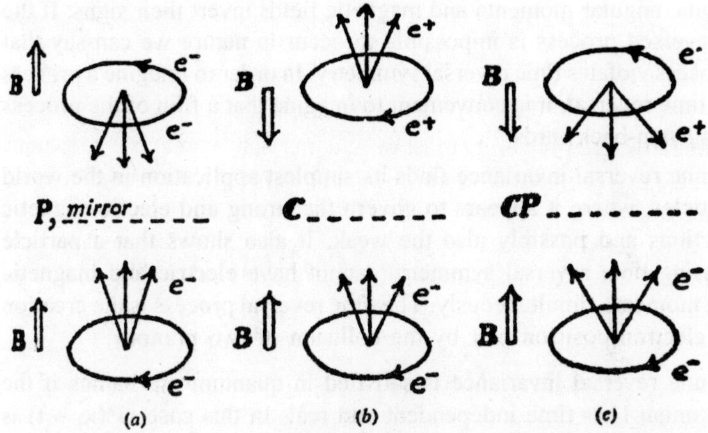

Fig. 7 : (a) Parity mirror, (b) charge conjugation mirror and (c) CP-mirror.

Let us consider the case of the β-decay of polarized nuclei (*e.g.* Co^{60}). The interpretation of the parity non-conservation, charge non-conservation and conservation under the combined operation. In this figure B shows the direction of a magnetic field due to current loop, used for polarizing the nuclei. It represents the nuclear spin and thus known as *polarization vector*. The upper diagrams represent the result of the reflection of the process space reflection creates a different system, as it changes the direction of decay arrows but not the direction of B. The charge conjugation only. This process leaves the direction of the decay unchanged, although electrons are replaced by positrons and the polarization direction is reversed. The combined effect of CP operation. This shows that under CP reflection, we obtain the process of decay of the anti-nucleus. From the above results we conclude that the reflection type of symmetry can be obtained by the combined operation of C and P only in weak interactions. Number of other examples show that the weak interactions do not grossly violate the combined CP-invariance. Unfortunately the decay of K-meson is not invariant under the combined operation CP.

7. Time Reversal

The time reversal operator is defined as that operator which reverses the direction of time, or the direction of all motions. Under this operation displacement, acceleration and electric fields remain invariant but momenta, angular momenta and magnetic fields invert their signs. If the time reversed process is impossible to occur in nature we can say that the process violates time reversal symmetry. In order to imagine a process under time reversal, it is convenient to imagine that a film of the process is being non-backwards.

Time reversal invariance finds its simplest application in the world of particles, where it appears to govern the strong and electromagnetic interactions and possibly also the weak. It also shows that a particle possessing time reversal symmetry cannot have electric and magnetic dipole moments simultaneously. The time reversal process is the creation of an electron-positron pair by the collision of two photons.

Time reversal invariance is satisfied in quantum mechanics if the Hamiltonian H is time independent and real. In this case $\psi^*(x, -t)$ is the time reversal wave function of $\psi(x, t)$. Thus time reversal operation T changes ψ as

$$T\,\psi(x, t) = \psi^*(x, -t). \qquad ...(18)$$

The motion of a particle is an external fixed magnetic field is not invariant under inversion of time. The relativistic treatment of time reversal shows that the inversion of time axis inverts the sign of the electrostatic potential. The (π^o-) mesic field, like the magnetostatic potential, is odd under time reversal in order to ensure that the interaction is time reversible.

8. Combined Inversion of CPT

The strong and the electromagnetic interactions are invariant under the separate operations of C, P, and T. The weak interaction does not conserve parity and also is not invariant under charge conjugation. All the interactions are invariant under the combined *strong reflection* operation CPT, irrespective of the order of the operations. No example of a violation of the CPT theorem is known. It follows that if T invariance is satisfied for all interactions, then these interactions will also be invariant under the combined operation of CP. The existence of CP violating interactions means that the analysis is not quite accurate.

The quantities conserved have the quantum numbers, which behave in two different ways, when one considers a system formed by the combination of two other systems. The quantum numbers, such as *angular momentum, isospin, strangeness, baryon number, lepton number, electric charge* are called *additive.* On the other hand, quantum numbers, such as *parity, invariance under charge conjugation, invariance under time reversal* are called *multiplicative.*

Let us now discuss the production and properties of elementary particles.

ELECTRONS AND POSITRONS

The electron is the first and best known elementary particle. It was discovered in 1897 by *J J. Thomson*. The mass of the electron at rest is 9.1091×10^{-31} kg. All other elementary particles have larger mass than that of the electron. Thus it is convenient to use the *mass of the electron as a wait for a mass.* The electron has a negative electric charge of magnitude 1.6021×10^{-19} coulomb and a spin 1/2. Its spin can still have either one or the other of the two possible conditions, usually known as *spin up or spin down.* According to the Dirac quantum mechanical theory of the electron the magnetic dipole moment associated with the electron spin should be exactly one Both magneton ($eh/4\pi m_e$). Another prediction of this theory is that the $2S_{1/2}$ and $2P_{1/2}$ atomic levels of an one electron atom should be degenerative. Both these predictions are wrong, the magnetic moment is about 0.1 per cent greater than the Dirac value and the leaves have a small energy difference.

In *1930, C.D. Anderson* in the United States and *P.M.S. Blackett* in England observed a particle corresponding to the electron, but having positive charge. This particle was called positive electron or *positron.* Just previously *P.A.M. Dirac*, in England, has predicted the existence of a positron in his relativistic theory of the free electron. Dirac suggested that electron positron pair cannot be created by bombarding particles or photons of energy less than the threshold energy $2mc^2$ (1.02 MeV). At the time of passage through ordinary matter, the positron may join an electron to form a positron-electron system, called *positronium,* which lasts for a measurable time before combining to produce *annihilation* radiation. Because of this annihilation property, the *positron is said to be the anti-particle to the electron.*

The lowest Bohr orbit of positronium is one for which n = 1 and $l = 0$ (S-State). This state has fine structure owing to the spins of the particles. The atom is in a ^1S state when the two spins are oppositely directed and in a ^3S State when the spins are parallel. Theoretical calculations by J. Pirenne (1944) and J.A. Wheeler (1946) showed that the lifetime of the singlet should be of the order of 1.3×10^{-10}. The annihilation radiation emitted by the combining of a positron-electron pair in the singlet state should consist of two gamma ray photons emitted simultaneously in opposite directions. Theoretical calculations by Ore and Powell (1949) showed that the lifetime of the triplet state should be about 1.5×10^{-7} sec. The annihilation should consist of three gamma ray photons travelling away from their origin at mutual angles of 120. We can explain the production of two in one case and of three γ-ray photons in the other as the result of the conservation of spin in the annihilation process. In the singlet state net spin in zero, two photons (spins +1 and −1) are expected in opposite directions in order to satisfy the law of conservation of momentum. In the triplet state (spin +1 or −1), three photons (+1, +1, −1 or −1, −1, +1) are expected. M. Deutsch (1951) measured experimentally the time between the formation of positrons and their subsequent annihilation and found the value 1.5×10^{-7} sec. This is the experimental evidence for the existence of the triplet state (*ortho*) of positronium.

PROTONS AND ANTI-PROTONS

Since early in the present century, the proton has been recognized as an essential component of matter. It is readily available for study and is the nucleus of the common hydrogen atom *(singly charged positive hydrogen atom H^+)*. The name proton, was suggested by Rutherford. Since the weight of the hydrogen atom is 1837 m, hence the weight of the proton is 1836 m_e. The charge of the proton is positive and exactly equal in magnitude to that of the electron. It is a fermion of intrinsic spin half. It is useful to regard the proton, as a particular charge state $T_3 = + 1/2$ of a single particle, the nucleon, of total isospin T = 1/2. Proton has a magnetic moment of 2.79275 nuclear magneton ($\mu_n = \mu.eh/4\pi m_p$).

Like the electron, the proton has a counter part of opposite charge with which it undergoes annihilation. Just as positron-electron pairs are formed by the interaction with matter of rays or particles having sufficient energy (1.02 MeV), so it should be possible to produce proton-anti-proton pairs in a similar manner, but it would require about 2000 times energy. Until 1955, the required energy was available only in high energy particles in cosmic radiations and the evidence of the antiprotons could be obtained from cosmic ray events in photographic emulsions and cloud chambers. With the completion of the 6 BeV bevatron at the University of California it was possible to perform an unambiguous experiment for the production and detection of anti-protons. The first convincing evidence of the anti-proton came from the experiment performed by Chamberlain, Segre Weigand and Yspsilantis in 1955. The schematic diagram of their detecting apparatus, which is as a mass spectrograph, is shown in Fig. 6. Protons of 6 BeV energy are incident on a copper target T. Anti-protons are produced by collisions between the incoming protons and protons in the target nuclei ($p+p \rightarrow 3p+p^-$). Negatively charged particles were selected by a mass spectrograph consisting of a series of deflecting magnets, focusing magnets, scintillation counters and Cerenkov counters. This selected beam of particles was expected to about 50,000 negative pions to every anti-proton. These particles all have the same momentum (1.19 BeV/c) and thus travel in the same trajectory in a magnetic field. Because of its larger mass the proton had a velocity of only 0.78c while the pion had a velocity of 0.99c.

Following arrangements were made for the identification of the anti-protons. The distance between plastic scintillation counters S_1 and S_2 was 40 ft.; the time of flight for anti-protons was 51 n sec. and for pions 40 n sec. Thus we see that the pulse emitted from counter S_2 occurs 51 n sec later than that from S_1. To obtain a coincidence the signal from S_1 was delayed enough that the two signals arrived at the coincidence circuit at exactly the same time. Sometimes there would be two different mesons passing through the two counters in such a way that a pulse from the second counter would be produced just 51 n sec. after a pulse from the first. This gave accidental coincidences appearing like anti-protons. To eliminate this difficulty the experimenters introduced the use of *Cerenkov counters*. The Cerenkov counter $\bar{C}_1$, of chemical composition $C_8F_{16}O$ and refractive index 1.28, was designed to respond to particles with velocities greater than 0.79 c. Any pulse from C_1 cannot be due to an anti-proton having velocity 0.78c. The second Cerenkov counter C_2, of fused quartz and refractive index 1.46, was designed to respond only to particle whose velocities were between 0.75c and 0.78 c. It was converted in anticoincidence with C_1, so that the pulses in C, were those produced by anti-protons only and in C_1 by negative pions only. The third scintillation counter S_a certifies that an anti-proton has not been scattered through a large angle in C. The distance between S_2 and S_3 was only a few inches.

In the experiment performed by Hill, Johanson and Gardner, the beam of 6.2 BeV protons were incident on a photographic emulsion. The proton produced a star (star A in Fig. 7) in a collision with a nucleus. One of the particles produced in this event was an anti-proton, which proceeded in the emulsion and found to give another star, whose several prongs were corresponding to particles appeared in this process. Measurements showed that the total energy released was greater than expected from the mass of the anti-proton alone, so that another particle, either a proton or neutron must have been annihilated simultaneously.

The annihilation of an anti-proton with a proton or neutron differs from that of a positron and electron in that several charged and neutral n-mesons together with protons and occasionally (if sufficient kinetic energy is supplied) deaterons and K-mesons are usually produced, rather than two or three quanta. The mass of the anti-proton has not been measured very accurately but it is presumably exactly equal to that of the proton. *Its electric charge is equal in magnitude, but negative in sign*, confirmed by the fact that protons annihilated with anti-protons. The energy released in proton antiproton annihilation is 1836.12 times the energy released in electron-proton annihilation.

NEUTRONS

Chadwick in 1932 discovered the neutron through a careful study of the kinematics of α-particle induced reactions with light nuclei. The production and properties have already been given in chapter.

In the series of experiments-which led to the detection of the anti-proton, the anti-neutron was also found. A convincing experimental evidence for the existence of anti-neutrons was announced in September 1956 by B. Cork, G.R. Lambertson and others. When a proton and anti-proton meet, mutual annihilation generally takes place. But if these two particles come fairly close to each other, although not doss enough for annihilation to occur, it is possible for an electrical charge to be transferred from the positive proton to the negative, or vice-versa. As a result, both particles would become electrically neutral, the proton being converted into a neutron and the anti-proton into an anti-neutron.

$$\overline{p} + p \rightarrow \overline{n} + n.$$

Evidence for this charge transfer process was obtained by allowing anti-protons, obtained from the Berkeley Bevatron, to enter a vessel of liquid hydrogen. About three out of every thousand antiprotons interacted With the protons in the hydrogen to form anti-neutrons. The latter were identified by the energy produced in the mutual annihilation of an anti-neutron and a neutron. The release of energy was accompanied by a flash of light in a detector. The recording system was designed to respond only to those associated with the correct amount of energy to be expected from neutron-anti-neutron annihilation. Since neither neutrons nor anti-neutrons have a net electric charge, hence the distinction of a particle from its anti-particle by its charge cannot apply to the neutron and anti-neutron. We do expect, however, that whatever be the internal charge distribution of the neutron, the anti-neutron will have the opposite internal charge distribution.

NEUTRIONS AND ANTI-NEUTRIONS

As discussed in chapter, it is possible to avoid violations of the law of conservation of energy, linear and angular momenta and spin by assuming the existence of a particle, called a *neutrino*. It has *charge zero, spin 1/2, zero rest mass (rest mass very small in comparison with that of the electron) and magnetic moment smaller than 10^{-8} Bohr magneton or nearly zero*. It also has energy and momentum.

Indeed, it has an anti-particle, the *anti-neutrino*, which is distinguishably different from it. The neutrino and anti-neutrino differ with respect to the property known as *helicity*. It is customary to call the particle accompanying a positron a neutrino, ν, while that accompanying an electron is called an anti-neutrino, $\overline{\nu}$. It is important to remember that positrons and electrons are never ejected alone. It is clear from experiments that neutrinos emitted in β-decay (positron decay and K-capture) have a negative helicity, i.e. they are longitudinally polarized with their spin axes anti-parallel to their direction of travel

$$H = +1 \text{ for } \overline{\nu};\ H = -1 \text{ for } \nu.$$

Because of its lack of charge and magnetic moment, a neutrino has essentially no interaction with matter except that which leads to inverse β decay. This interaction is extremely weak.

PHOTONS

The photon is the quantum of the electromagnetic field. It was first recognized by Einstein in 1905, who extended Planck's quantum hypothesis to apply to radiation in order to explain certain puzzling features of the photoelectric effect. The neutrino and the photon have a number of identical properties. The main difference in their behaviour springs from the different spin values. The spin is half for the neutrino and one for the photons. Similar to neutrino, photon has no mass and no electric charge.

The spin of the photon is associated with its polarization. Bethe measured the torque communicated to a suspended quartz disc of a torsion balance experiment illuminated with polarized light. From his measurements he was able to show that each photon polarized light carried an angular momentum of magnitude $\hbar$.

Left and right circularly polarized light corresponds to opposite senses of rotation of the photon. A beam in which the electric vector at a given point rotates anti-clockwise when viewed from the light source is known as a *left circularly polarized light beam*. A left circularly polarized photon is the anti-panicle of a right circularly polarized photon. *Actually photon and the anti-photon are one and the same, there is no distinction between them*. The above classification refers only to plane polarized light which is conveniently regarded as a mixture of left and right circular polarized components.

Mesons

In the summer of 1935, C.D. Anderson and S.H. Neddermyer observed few tracks, which could not be attributed to either protons or electrons, in cloud chamber cosmic ray observations. They concluded that the new particle might be ligher than a proton and heavier than an electron or positron. Almost simultaneously and independently J. C. Street and E. C. Stevenson also obtained a cloud chamber photograph and found a negatively charged particle of mass ~ 130 m_e. During 1937–38, confirmation was obtained from various sources of the existence in cosmic rays of particles lighter than a proton but about 200 times heavier than the electron. Such particles were found to carry either positive or negative charge and named *mesotrons* by Anderson and Neddermyer, from the Greek word *meso*, meaning intermediate. Later, the name meson, suggested by the Indian physicist *H. J. Bhaha* in 1939, was adopted. The existence of this type of particle had actually been predicted by Yukawa in 1945. In his study of the nuclear force field Yukawa predicted quantum mechanically the existence of a system of nucleon exchange forces which required the existence of a particle, with either positive or negative charge and of mass equivalent to 200 or 300 m_e, in order to give the correct distance over which the short range forces act. Before 1947 there was no evidence for the existence of more than one type of intermediate particle.

In the same year Powell, Occhialini and others in England showed that two kinds of mesons with different masses existed. Upon developing photographic films with nuclear track emulsions after exposure for some time at high altitudes in the Bolivian Andes, tracks were observed indicating the presence of a meson with a mass ~ 300 m_e. It was also noted that at the end of its path in the emulsion, the heavier meson changed into a lighter particle with a mass of about 200 m_e. These two mesons were distinguished by calling the heavier one the π-meson and the lighter the μ-meson.

According to the new definition of mesons, they must have a spin quantum number that is either zero or integral. As *μ-mesons have spin half, so that it is not a meson in the correct sense of the term*. π-mesons are commonly referred to as muons.

The π-mesons, also called pions, are not only heavier but 100 times more stable than muons. The π-meson (primary meson as it decays to muon $\pi^{\pm} \rightarrow \mu^{\pm} + \nu$) is identifiable as the Yukawa particle, but the muon is a puzzle : *nobody should be different from that of an electron*, since muons and electrons are identical in all other ways. This particle is still unexplained and unneeded.

MUON

Production of Muons By bombarding graphite with accelerated α-particles of energy 450 MeV it was found that mesons could be created artificially. The charge of muons is equal in magnitude to the charge of the electron. There are no neutral muons. They are usually obtained as a decay product of the pion according to the reactions:

$$\pi^+ \rightarrow \mu^+ + \nu_\mu$$

$$\pi^- \rightarrow \mu^- + \nu_\mu.$$

The neutrino appearing in this reaction seems to be different from the neutrino of β-decay. One occasionally gets muons at 1 BeV which are interpreted as the product of the reactions.

$$\nu_\mu + n \rightarrow \mu^- + p$$

$$\overline{\nu}_\mu + \rightarrow \mu^+ + n,$$

However, we know reactions

$$n_e + n \rightarrow e^- + p$$

$$\overline{\nu}_e + p \rightarrow e^+ + n.$$

Under the conditions of the bombardment one would also expect to produce electrons, *provided the two kinds of neutrinos are identical*. In fact, no electrons have been found, resulting, that *there are two different kinds of neutrinos*.

Although, most common source of muons is the pions decay but they also appear as branchings in other decays such as

$$\Sigma^{\pm} \rightarrow \mu^{\pm} + \nu + n$$

$$\Delta^{o} \rightarrow \mu^{-} + \nu + p$$

$$K \rightarrow \mu + \nu, \text{ etc.}$$

Muons can also be generated in pairs by photo production, similar to electron pair production.

Mass of the Muons

Early determinations of the mass of the muons in cosmic rays were made by utilizing the tracks in a cloud chamber in a magnetic field or in a photographic emulsion. The quantities measured were momentum and energy. From these one finds the mass. Barkas and co-workers (1956) measured the ratio of the pion-muon mass and the pion-proton mass, by measuring momenta and ranges in photographic emulsion. These measurements gave the masses with an accuracy of one part in a thousand.

A complexly different way of obtaining the muon mass is offered by the study of relative transitions in μ-mesonic atoms. In the last few quantum jumps from one orbit to another, muonic X-rays are radiated. For high Z-absorbers, the ^1S muon orbits penerate the nucleus and the finite size of the nucleus strongly affects the energies of the radiated X-rays. For low Z elements, the M and L muonic X-rays have energies approximately the same as given by Bohr theory. In 1960, Devons and coworkers have determined the muon mass by measuring the mass absorption co-efficient of the $^3D_{5/2} \rightarrow {}^2P_{3/2}$ muon transition of phosphorus in lead. A negative muon brought to rest in the phosphorus is recorded by a coincidence between pulses from counters 1, 2, 3, 4 and an anti-coincidence in counter 5. A muonic X-rays photon is recorded by a NaI detector D placed behind a lead absorber. For phosphorus (Z = 15), the strongest line $D_{5/2} \rightarrow P_{3/2}$ has an energy

$$E({}^3D_{5/2} \rightarrow {}^2P_{3/2}) = 88{,}000 \pm 10 \text{ eV}.$$

According to the Bohr theory

$$E = hc\, R\left(\frac{1}{2^2} - \frac{1}{3^2}\right) Z^2 (1.001163) \frac{m_\mu}{m_e} \qquad ...(19)$$

This relation gives $m_\mu = 206.78\ m_e$. In this method we have assumed that the charge of the muon is exactly equal to that of the electron.

Recent accurate measurements of the gyromagnetic ratio and of the magnetic moment of the muon provide $m_\mu\ (206.77 \pm 0.01)\ m_e$.

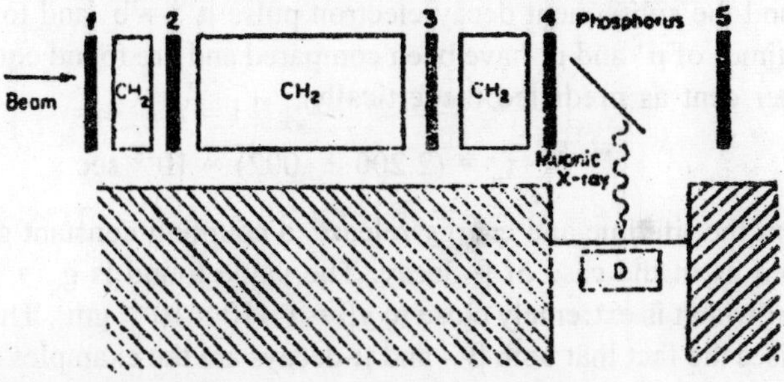

Fig. 8 : Devons experiment for the determination or negative muon rest mass.

Muon Decay and Mean Life

The decay of muon has been observed in cloud chamber photographs and in nuclear photographic emulsions. In 1949, Leighton, Anderson and Seriff measured the energies of the electrons and positrons produced in the decay of cosmic ray γ-mesons at sea level.

They placed Geiger counter controlled cloud chamber in a magnetic field of 7200 gausses and found a range of values from 9 to 55 MeV for the decay electrons. This wide range of energies for the decay electron spectrum suggests that this decay process is accompanied by more than one invisible neutral particle to conserve energy and momentum. The form of the spectrum agrees with the predictions of reactions —

$$\mu^- \rightarrow e^- + \nu_\mu + \overline{\nu}_e$$

$$\mu^+ \rightarrow e^+ + \overline{\nu}_\mu + \nu_e.$$

Of the two neutrinos, one is believed to be of the kind found in beta decay and the other of the kind found in pion decay. Here the choice of $\nu_e + \overline{\nu}_\mu$ is preferred because it associates the e^+ with the ν_e as in beta decay.

For electrons having the maximum possible energy, the neutrino pair has zero angular momentum in that direction and the electrons must have the same angular momentum as the decaying muon. The decay of μ^- would require that if e^- is defined as a particle, μ^- must likewise be a particle; e^+ and μ^+ are then anti-particles.

The mean life of the free muons was originally informed from their disappearance in cosmic rays. The most recent work uses the signal from a muon production process $\pi^+ \rightarrow \mu^+$ in a scintillator to start a timing circuit and the subsequent decay electron pulse $\mu^+ \rightarrow e^+$ and to spot it. The lifetimes of μ^+ and μ^- have been compared and are found equal, also to 0.1 per cent as predicted theoertically.

$$\tau_\mu = (2.200 \pm .002) \times 10^{-6} \text{ sec.} \qquad ...(20)$$

From the lifetime and energy release, a coupling constant g can be calculated as in the case of β-decay. The value found is $g_\mu = 1.435 \times 10^{-62}$ Jm^3 which is extremely close to $g\beta = 1.410 \times 10^{-62}$ gm^3. This result emphasises the fact that both β – and μ-decays are the examples of weak interaction.

Interaction with Matter

Negative muon captured into Bohr orbit surrounding nuclei may undergo nuclear capture

$$\mu^- + p \rightarrow n + \nu,$$

instead of radioactive decay, analogous to orbital electron capture. The probability of muon capture by a nucleus increases with the atomic number Z of the nucleus. The probabilities of capture and free decay becoming equal when the .atomic number Z is about 12. Consequently, in matter having Z less than 12, most of the negative muons will decay with the normal muon life time, but for Z greater than this value they are captured before they can decay.

The passage of muons through matter is analogous to that of electrons, except that the decay processes decrease with Z. Cosmic rays studies show that fast muons can produce nuclear disintegration stars, which can be interpreted as an electromagnetic process. Similar to the electron scattering, muon beams of energy upto several hundred MeV have been used for nuclear scattering experiments. Slow μ^- from muonic atoms, slow μ^+ may from muonium. The μ^- may also bind two atoms to form muonic molecule. Pair production of muons (μ^+, μ^-) by high energy bremsstrahlung has also been observed.

Spin and Magnetic Moment of Muons

The experimental measurement of the spin and of the magnetic moment of the muon is intimately connected with the discovery of the non-conservation of parity in muon decay. Assume that a pion comes to rest and decays, emitting a neutrino and a muon in opposite directions. Since the pion has no angular momentum, hence the muon has an angular momentum equal and opposite to that of the neutrino. If the neutrino is polarized with the spin anti-parallel to the direction of motion, so is the muon.

This has been verified by direct experiment. If the muons are brought to rest in a target medium, which does not seriously depolarize the muons, we have the emission of a positron (for μ^+ mesons) and two particles of zero rest mass from a polarized source. The positrons tend to travel parallel to the spin direction. Now if we apply magnetic, field to the muon at rest, the field produces a rotation of the direction of polarization with an angular velocity (Larmor frequency)

$$\omega = g\left(\frac{eB}{2m_{\mu}}\right)2\pi\,\mu_0, \qquad ...(21)$$

where g (= μ/I) is the gyromagnetic ratio, i.e. the magnetic moment in natural $\mu_0 e\hbar/2m_\mu$ divided by angular momentum in units of $\hbar$ By using these phenomena Friedman and Telegdi performed experiments and found $g = 2 \times (1.0020 \pm 0.0005)$, very close to the g value of the electron. This result is a strong argument for a spin 1/2 of the muon.

The latest and most precise experiment is due to Bailey, Parley, *et al.* This compares the Larmor frequency of the muon in a given magnetic field with the orbital frequency of the particle in the same field. These two are equal, if g = 2. For a finite value of g – 2, a beam of longitudinally polarized muons in a circular orbit will gradually build up misalignment between the spin direction and the direction of motion. The use is made of a strong ring, filled with fast onions produced by pion decay in the same orbit. The muons decay in the flight and the detector circuits are set to respond only the forward going electrons having the highest energy. Since the angular distribution of emission from muon decay is not isotropic, the observed intensity of electron emission is modulated by the muon precession due to the anomalous moment. Counting rate as a function of time after the injection of pious into the ring gives

$$a = 1/2\ (g' - 2) = (116616 \pm 31) \times 10^{-8} \qquad ...(22)$$

for μ^+ and μ^- both.

PIONS

The result of cosmic ray investigations led to the conclusion that pions are produced both in nucleon-nucleon interaction and in the interaction of high energy photons with nucleons. It is clear from this conclusion that pions can be produced in laboratory by the use of high energy particles obtained by means of an accelerator. The amount of energy which the accelerated particle must have depends on the mass of particle and the nature of the target but it will be always greater then the rest mass of the pion.

Artificial Production

The pions were first produced artificially by Gardner and Lattes in 1948, using a beam of 380 MeV α-particles from the Berkeley Synchrocyclotron. This beam was allowed to fall on the edge of a carbon target and the mesons emitted were detected in nuclear emulsions. For laboratory production of π-mesons, the most obvious target materials are hydrogen and deuterium or substances rich in these isotopes. At the present time, strong beams of charged pions for experimental purposes, with energies covering a wide range, are produced by bombarding solid targets, *e.g.*, Be, with accelerated protons.

The simplest nucleon-nucleon interaction involving protons as projectile and target and producing pions is evidently.

$$p + p \rightarrow \pi^+ + d$$

or $$p + p \rightarrow (n + p).$$

The interaction between a high energy proton and a deuteron may be considered as an interaction between the proton and one of the nucleons which constitute the deuteron. The possible reactions are.

$$p + p\ (\text{in d}) \rightarrow \pi^+ + P + n$$

$$p + p\ (\text{in d}) \rightarrow \pi^+ + d$$

$$p + n(\text{ind}) \rightarrow \pi^+ + n + n$$

$$p + n(\text{ind}) \rightarrow \pi^- + p + p.$$

The production of pions by gamma radiation incident on nucleons has been studied extensively in the energy range from threshold to 450 MeV and less thoroughly upto 1 BeV.

$$\gamma + p \rightarrow \pi^+ + h$$

$$\gamma + n \rightarrow \pi^- + p.$$

Besides charged pions, neutral pion has been produced by bombarding hydrogen and deuterium with high energy photons.

$$\gamma + p \rightarrow \pi^o + p,\ \gamma + d \rightarrow \pi^o + (n + p),\ \gamma + d \rightarrow \pi^o + d$$

and by proton-proton interaction as

$$p + p \rightarrow \pi^o + p + p.$$

The source of gamma rays is generally the bremsstrahlung of electrons accelerated by machines. The-production of neutral pions by the interaction of high energy photons with matter was first established by Steinberger, Panofsky and Steller in 1950. In their experimental arrangement a narrow beam of 330 MeV X-rays from the synchrotron was incident on a target (beryllium or hydrogen) and the products were detected by two gamma ray counters making an angle f with each other in a plane at an angle θ with the X-ray beam. Each counter consisted of three crystal scintillation counters with a thin piece of lead between the first two crystals to convert the gamma ray photons into electron-positron pairs. The measured values of the coincidence rates for different values of ϕ and θ agreed with the calculated values based on the assumption that a π^o meson was disintegrated into two gamma ray photons.

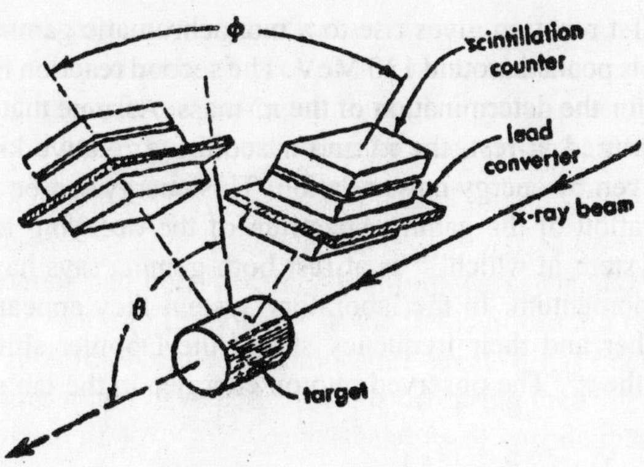

Fig. 9 : Experimental arrangement for the production of π^o-meson.

Pions are commonly formed in the decay of kaons, hyperons and resonant states. Since there is no requirement for the conservation of pions, there is no limitation on their formation even in strong interaction.

Mass of Pions

The mass of the charged pion can be determined by methods similar to those used for muons. The rest mass of the negative pion can be found by comparing the observed pionic X-ray energies for the 4f-3d orbit transitions in media of low atomic number ($Z < 20$) with the theoretical expression which contains m as an unknown parameter. Apart from the reduced mass, the only important correction to the Bohr formula is the effect of vacuum polarization (~ 0.2 to 0.3). The rest mass of the negative pion thus obtained is

$$m_\pi = (273.2 \pm 0.1)m_e = 139.577 \pm 0.013 \text{ MeV}.$$

The positive pions have the same mass and can be determined by the same method. The mass of the neutral pion is found by a quite different method. When very low energy negative pions are slowed down in hydrogen they are captured into orbit. These cascade down the optical levels and are ultimately captured by the proton. The reactions observed are:

$$\pi^- + p \rightarrow n + \gamma$$

$$\pi^- + p \rightarrow n + \pi^o$$

$$\pi^o \rightarrow \gamma + \gamma.$$

The first reaction gives rise to a monochromatic gamma ray whose lab energy is peaked around 130 MeV. The second reaction is particularly important for the determination of the π^o-mass. Assume that the negative pion is captured at rest, the π^o and n acquire a relative kinetic energy equal to given by energy mass relation. This energy can be measured by the observation of the gamma spectrum of the decaying π^o. In the co-ordinate system in which π^o is at-rest both gamma rays have equal and opposite momentum. In the laboratory system they appear at an angle to each other and their frequency shows the Doppler shift due to the motion of the π^o. The observed photon energies in the lab system range from

$$E_{\gamma\,(min)} = \frac{1}{2}\, m\pi^o c^2\, (1 - \beta)\, (1 - \beta^2)^{-1/2} \qquad ...(23)$$

to

$$E_{\gamma\,(max)} = \frac{1}{2}\, m\pi^o c^2\, (1 + \beta)\, (1 + \beta^2)^{-1/2}. \qquad ...(24)$$

These limits are obtained from relativistic Doppler formula for a π^o decaying at the instant; speed is βc. It is easy to write the rest mass energy difference as

$$(m_{\pi^-} - m_{\pi^o})\, c^2 = (m_n - m_p)\, c^2 + \frac{\alpha^2 c^2 m_{\pi^-}}{2} + \frac{\frac{1}{2}\left(m_{p} - m_{\pi^-}\right)\Delta E_{\gamma^2}}{\left[m_\pi - \left(m_p - m_p\right)\right] m_n c^2} + \ldots$$

Here the second term on the right is the pion binding energy in a ^1S Bohr orbit and the third term is the kinetic energy of the π^o. In the second term a is the fine structure constant. Using known values of $\Delta E_\gamma (= E_{\gamma,\,max} - E_{\gamma,\,min} = p_{\pi^o} c)$ α, m_n, m_p and $m\pi^-$ we have

$$\left(m_{\pi^-} - m_{\pi^o}\right)c^2 = 4.6 \text{ MeV}$$

or

$$m_{\pi^o c^2} - m_{\pi^- c^2} - 4.6 \text{ MeV}$$

$$= 139.577 \pm 0.013 - 4.6 = 134.977 \pm 0.013 \text{ MeV}.$$

Decays

Unless they have sufficiently high energies (~ 150 MeV), positive pions do not interact to any great extent with nuclei and decay at rest according to the reaction.

$$\pi^o \rightarrow \mu^+ + \nu_\mu.$$

Since positive muon is an anti-particle, the ν_μ must be a neutrino to conserve the muonic lepton number. This decay is due so the weak interaction and the parity is not conserved. The onion is emitted with a unique energy of 4.17 MeV and is longitudinally polarized. This energy is consistent with zero mass for the muon associated neutrino. Pions may also decay according to the reactions

$$\pi^+ \rightarrow e^+ + \nu_e \,(1.24 \times 10^{-2}\%)$$

$$\pi^+ \rightarrow \mu^+ + \nu_\mu + \gamma \,(1.24 \times 10^{-2}\%)$$

$$\pi^+ \rightarrow \pi^o\, e^+ + \nu_e \,(1.5 \times 10^{-6}\%)$$

The first equation requires that the helicity of the positron shall be the same as that of the neutrino. The third equation is analogous to a $0^+ \rightarrow 0^+$ Fermi transition in nuclear β-decay.

Similar decay processes are possible with negative pions.

$$\pi^- \rightarrow \mu^- + \overline{\nu}_\mu, \pi^- \rightarrow e^- + \overline{\nu}_e \text{ and } \pi^- \rightarrow +e^- + \overline{\nu}_e$$

Negative pion decay ($\pi^- \rightarrow \mu^-$) is not observed at rest because of the Coulomb attraction with the nuclei and strong nuclear absorption.

The decay modes of the π^o are

$$\pi^o \rightarrow \gamma + \gamma \qquad (98.8\%)$$

$$\pi^o \rightarrow \gamma + e^+ + e^- \qquad (1.2\%)$$

$$\pi^o \rightarrow e^+ + e^- + e^+ + e^- \quad (0.003\%)$$

It is evident from the nature of the products that π^o-mesons decay by the electromagnetic interaction while charged pions decay by strong and the weak interactions both.

Mean Life

One of the earliest determinations of the mean lifetime of charged pions was made by J.R. Richardson (1948), using the π-mesons produced by the bombardment of graphite by 380 MeV α-particles.

These pions travelled in helical paths in the vertical magnetic field of the cyclotron. Those which rose upward were allowed to travel through 180° before striking a photographic plate, those which spiraled downward were allowed to travel 540° before striking a photographic plate.

The number of mesons which decayed while traversing a complete cycle can be determined by the ratio of the number of mesons in the two plates. The time required to traverse a circle in a cyclotron is independent of the speed, hence the average life time of the decaying mesons could be determined. The first result for the mean life of charged pion obtained was 1.1×10^{-8} sec.

A more recent experiment of the decay in flight of positive pions was performed by Durbin, Loar and Havens in 1952. The meson beam produced in the Nevis cyclotron by the bombardment of a beryllium target with 385 MeV protons was deflected by a magnet and directed through three stilbene scintillation counters operated in coincidence.

A much larger scintillation counter, C_4, 8" dia and containing xylene, was set at a distance L which could be varied from a few inches to about 8 ft. The coincidence counters C_1, C_2 and C_3 defined the beam of the π-mesons.

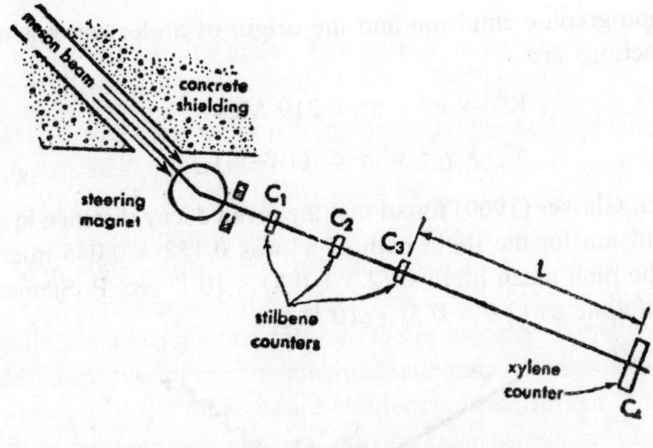

Fig. 10 : Measurement of life time of π-mesons.

Some of these mesons decayed in flight between C_3 and C_4, the remaining ones being detected by C_4. After suitable corrections for beam impurities, angular divergence, accidental coincidences, the difference in counts between C_1, C_2, C_3 detectors and the C_4 detector represented the number of pions decaying in the flight along the path of length L. Knowing the average velocity and determining mean free path, the mean life time could the calculated. The value so obtained was 2.54×10^{-8} sec for π^+ mesons and 2.44×10^{-8} sec. for π^- mesons. Thus within the limits of experimental error, the mean lifetimes of the π^+ and π^- mesons were identical, with the average mean life = 2.5×10^{-8} sec.

Using a beam of 34 MeV positive pions from the Rochester 130" cyclotron, Kinsey, *et al.* (1964) found pion lifetime as $(2.640 \pm 0.008) \times 10^{-8}$ sec. By observation of decay in flight, Ayres et al; (1968) found that the lifetimes of the π^+ and π^- mesons are equal within standard deviations and suggested the value 2.6×10^{-8} sec.

The decay of π^o goes through a virtual intermediate nucleon- anti-nucleon state.

$$\pi^o \xrightarrow[\text{Strong}]{} \underset{\text{Virtual}}{(n+\bar{n})} \xrightarrow[\text{EM}]{} \gamma + \gamma .$$

This two stage decay involves both strong and electromagnetic interactions, hence suggests a short mean life of the π^o. The accurate method of measuring the mean lifetime of π^o-mesons is to mesure the mean distance between the end of a track of a K^+ meson brought to rest in a photographic emulsion and the origin of a electron positron pair. The reactions are

$$K^{o} \rightarrow \pi^{+} + \pi^{o} + 219 \text{ MeV}$$

$$\pi^{o} \rightarrow \gamma + e^{+} + e^{-} \ (1.24\%)$$

R.G. Glasser (1960) found that the mean decay distance in an Ilford L-4 emulsion for the flight path of π^{o} was 0.122 ± 0.045 microns and hence the pion mean lifetime $(2.3 \pm 0.8) \times 10^{-16}$ sec. P. Stamer (1966) found lifetime as $(1.0 \pm 0.5) \times 10^{-16}$ sec.

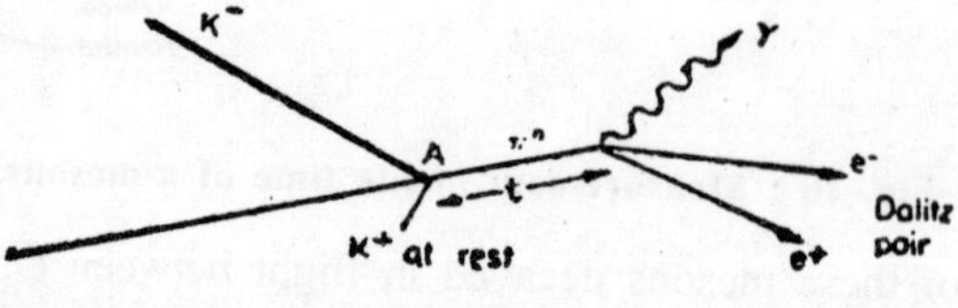

Fig. 11 : Track in photographic emulsion for the measurement of the mass π^{o}-mesons.

The lifetime of π^{o} has also been determined from the cross-section for the photo-production of π^{o} mesons in the Coulomb field of a heavy nucleus. The average of all observations is $(0.89 \pm 0.18) \times 10^{-16}$ sec. The mean decay distance in an Ilford L-4 emulsion for the flight path of the π^{o}-mesons was 0.122 ± 0.045 microns which leads an estimate for the mean lifetime of $(2.3 \pm 0.8) \times 10^{-16}$ sec.

INTERACTION WITH MATTER

Because of their short life time ($\sim 10^{-16}$ sec.), it π^{o}-mesons move only a few atomic diameters before they decay and thus are not affected by the matter through which they pass. Positive pions are not able to surmount the nuclear Coulomb barrier and, therefore, undergo spontaneous decay, while negative pions are captured by nuclei. All of the energy of a captured pion is transmitted to the nucleus results to a nuclear disruption, *called a star*, because of its appearance in a particle track chamber or photographic emulsion. Menon, Muirhead and Rochat in 1950 investigated the stars produced by the capture of slow π^{-}-measons in the nuclei of the elements composing nuclear emulsions. They studied more than 2500 stars and classified according to the number of prongs in each. About 46% of all visible stars were ascribed to the interaction of π^{-}-mesons with light nuclei such as C^{12}, N^{14}, O^{16} and about 54% were with the heavy nuclei, such as silver and bromine. All the π^o-mesons absorbed in the emulsion do not produce visible stars.

By using available charged meson beams and hydrogen targets we can measure three cross-sections, corresponding to the reactions:

$$\pi^+ + p \rightarrow \pi^+ + p,\ \pi^- + p \rightarrow \pi^- + p,\ \pi^- + p \rightarrow \pi^o + n.$$

The first two reactions are purely elastic scattering and the last reaction is known as charge exchange scattering. The first accurate measurements upto 200 MeV π-meson indicated that the total cross-sections of reactions were in the ratio of 9 : 1 : 2.

The scattering cross-sections can be interpreted by assuming that in the strong interactions the total isospin is conserved as well as the third component. The total isospin of the π^+, p system is 3/2 and the component T_3 = 1 + 1/2 + 3/2. The π^-, p system has T_3 = – 1 + 1/2 = – 1/2 and is a mixture of T = 3/2 and T = 1/2 states. If the scattering amplitudes corresponding to the scattering in the T = 3/2 state and the T = 1/2 states are a(3/2) and a(1/2) respectively, then the cross-sections are:

$$\sigma_+ \propto \left|a\left(\frac{3}{2}\right)\right|^2 \qquad ...(25)$$

$$\sigma_- \propto \frac{1}{9}\left|2a\left(\frac{1}{2}\right)+a\left(\frac{3}{2}\right)\right|^2 \qquad ...(26)$$

$$\sigma_0 \propto \frac{2}{9}\left|a\left(\frac{3}{2}\right)-a\left(\frac{1}{2}\right)\right|^2. \qquad ...(27)$$

If the π-p scattering predominantly occurs in the T = 3/2 state near 200 MeV, $a\left(\frac{3}{2}\right) >> a\left(\frac{1}{2}\right)$ and the cross-sections satisfy the ratios

$$\sigma_+ : \sigma_- : \sigma_0 = 9 : 1 : 2. \qquad ...(28)$$

From general collision theory, the maximum cross-section for elastic scattering is

$$\sigma_{max} = 4\pi\lambda\!\!\!^{-2} \frac{2J+1}{(2s_1+1)(2s_2+1)} \qquad ...(29)$$

In the case of π^+-p system s_1 the intrinsic spin of the pion is zero, s_2 the intrinsic spin of target proton is 1/2. For pions upto 300 MeV only s and p wave scattering is important, The d wave impact parameter is beyond the range of the *pion nucleon* force. The maximum value of total cross-section scattering of n π^+-mesons by protons was found 192 millibarns below approximately 200 MeV. At the peak of the resonance $\pi\lambda^2 = 24$ mb. These values with eqn (29) give us total angular momentum of the π-p system J = 3/2. This shows that the interaction is dominated by p-wave scattering

$$\left(L = J - s = \frac{3}{2} - \frac{1}{2} = 1\right).$$

Absorption of stopped π^--mesons : The fast charged pion slows down from an energy of about 100 MeV to about 1 keV in liquid hydrogen in 10^{-9} sec. In the further 10^{-12} sec, π-meson loses energy and becomes bound in a 1s orbit to a proton. From this orbit it may decay or react with the proton as

$$\pi^- + p \to \pi^o + n,\ \pi^o \to 2\gamma \qquad ...(30)$$

or

$$\pi^- + p \to n + \gamma. \qquad ...(31)$$

Panofsky, Aamodt and Hadley studied these reactions and found peak about 130 MeV describing reaction (31) and lower peak of mean energy of 70 MeV representing the energies of photons from the π^o-meson decaying as shown by reaction (30). The initial state has spin and parity $J = 1/2^-$, since the π^- is captured from K-shell and the pion has odd parity. In reaction (31), the neutron state has $J = 1/2^+$, the photon must, therefore, carry odd parity and unit angular momentum.

Intrinsic Spin

The reactions d + p = d + π^+ have been used to determine the spin of the π^+-meson (and therefore presumably of the π^--meson), using the measured cross-sections of the two reactions together with the *statistical mechanical principle of detailed balance.*

The differential cross-section for the reaction $p + p \to d + \pi^+$ given in terms of transition matrix elements is

$$\frac{d\sigma_{pp\to d\pi}}{d\Omega} = \frac{1}{(2s_p + 1)(2s_p + 1)}\left(\frac{1}{q_{pp}}\right)^2 \Sigma|T_{ba}|^2 \qquad ...(32)$$

where q_{pp} is the momentum in C.M. system of the initial state pp. Here T_{ba} is the matrix elements with a initial plane wave state and b the final state.

Similarly for the reaction d + $\pi^+ \rightarrow$ p + p, we have

$$\frac{d\sigma\pi_{d\rightarrow pp}}{d\Omega} = \frac{1}{(2s_d+1)(2s_\pi+1)}\left(\frac{1}{q\pi_d}\right)^2 \Sigma|T_{ab}|^2. \quad ...(33)$$

If both cross-sections refer to the same total energy and to the same angle of scattering in C.M. system, we contain the detailed balancing relation

$$\frac{d\sigma_{pp\rightarrow d\pi}}{d\Omega} = \frac{(2s_d+1)(2s_\pi+1)}{(2s_p+1)(2s_p+1)}\left(\frac{q_{\pi d}}{q_{pp}}\right)^2 \frac{d\sigma_{d\pi-pp}}{d\Omega}. \quad ...(34)$$

Using $S_p = 1/2$, $S_d = 1$ and the measured cross-sections and momenta in C.M. system. Cartwright, et al and Durkin, et al. suggested spin of the π^+-meson as zero.

It is assumed that the π-spin is the same as that of the π^+

The spin of the neutral pion cannot be determined directly. We know that π^o-meson decays into two photons and never into three photons. This implies that the intrinsic spin must be zero. Thus we see that all pions have spin zero and have no magnetic moment.

Parity

A negative pions are slowed down and captured into Bohr orbits in deuterium gas under pressure, we get following reactions :

$$\pi^- + d \rightarrow n + n \quad ...(35)$$

$$\pi^- + d \rightarrow n + n + \gamma \quad ...(36)$$

$$\pi^- + d \rightarrow n + n + \pi^o. \quad ...(37)$$

Reaction (37) is not observed. Panofsky measured the cross-section ratio

$$\frac{\sigma(\pi^- + d \rightarrow n + n)}{\sigma(\pi^- + d \rightarrow n + n + \gamma)} = 2.36 \pm 0.36 \quad ...(38)$$

Slow π^- mesons are absorbed by the nuclei of hydrogen as $\pi^- + p \rightarrow \pi^o + n^o$ and $\pi^- + p \rightarrow n + \gamma$. The ratio of these two reactions has been determined by different workers. The latest value of this ratio is 1.53 ± 0.02.

To interpret these two ratios, there are following assumptions : *(1) The π^--mesons are absorbed by the nucleus from s-states of the mesic atoms formed in the capture of the π^- by the atoms of deuterium and hydrogen. (2) The proton and neutron have the same parity.*

From the reactions (35) and (36), in order for the final states to be consistent with Pauli's Principle, two neutrons can exist only in $^1S_0$; $^3P_{0, 1, 2}$; $^1D_2$ or $^3F_{2, 3, 4}$ etc. states. The total angular momentum of the initial state, i.e. the pion and deuteron system, is just the spin of the deuteron, J = 1. The parity, given by $P_\pi K_d (-1)^l$ is equal to P_π since P_d is even and $l = 0$.

To conserve the total angular momentum and parity, the n + n system must be in a J = 1 state and must have same parity as that of the π^-meson. For L-even, the nucleon spins must be anti-parallel (S = 0) and for L-odd the nucleon spins must be parallel. Thus the only state with J = L + S = 1 is the $^3P_1$ state. This corresponds to an odd parity. Thus *the intrinsic parity of π-meson is negative.* This conclusion is consistent with the suppression of reaction (37), since π-meson also has negative parity. The similar reaction $\pi^+ d \rightarrow pp$ at rest cannot occur since Coulomb repulsion keeps if*- and d apart. A detailed analysis of the cross section suggests negative parity for π^+-mesons. The parity of the neutral pion is difficult to determine experimentally. It is again negative, can deduced using rare decay mode

$$\pi^+ \rightarrow e^+ + e^- + e^+ + e^-$$

and measuring distribution of angles between the planes of the two electron pairs.

K-MESONS

K-mesons are produced when high energy protons bombard a suitable target. The pions are also produced in this process. With increasing proton energy the fraction of K-mesons relative to pions increases. There are essentially four K-mesons, i.e. two charged mesons K^+ and K^-, which are particle and anti-particle of one another, and two uncharged mesons K° and K^{-o}, which are also particle and anti-particle. These are obtained in laboratory by the interaction of high energy protons or pions with various atomic nuclei.

Decay and Lifetime

Both neutral and charged K-mesons decay into ordinary particles. Assuming that $\Delta S = \pm 1$ in the decay, it is necessary to assign strangeness

S = + l or S = – l to the K-mesons. For S = + l, equation $Q = T_3 + B/2 + S/2$ implies that $T_3 = \pm 1/2$, corresponding to K^+ and K^o mesons. In addition, two corresponding anti-particles are expected for S = –l, these are K^- and $\overline{K^o}$ with $T_3 = -1/2$ and $+1/2$ respectively.

There are six different ways that K^+-mesons commonly decay, in each case giving two or three less massive particles. The various decay modes of the K^+ are denoted by symbols, θ^+, τ, τ', $K_{\mu\gamma}$, $K_{\mu g}$ and $K_{\mu 3}$. This confusing terminology dates from the time when it seemed likely that the θ^+, τ.... decays sprang from different parent particles. The notation $K_{\mu 3}$ represents a decay into three particles one of which is a muon. The decay reactions are :

$$K^+(\theta^+) \rightarrow \pi^+ + \pi^o + 219.3 \text{ MeV} \qquad (\sim 25\%)$$

$$K^+(\tau) \rightarrow \pi^+ + \pi^- + \pi^+ + 75.1 \text{ MeV} \qquad (\sim 6\%)$$

$$K^+(\tau') \rightarrow \pi^+ + \pi^o + \pi^o + 84.3 \text{ MeV} \qquad (2 \sim \%)$$

$$K^+(K_{\mu 2}) \rightarrow \mu^+ + \nu + 388.2 \text{ MeV} \qquad (\sim 58\%)$$

$$K^+(K_{\mu 3}) \rightarrow \mu^+ + \pi^o + \nu + 253.2 \text{ MeV} \qquad (\sim 5\%)$$

$$K^+(K_{e3}) \rightarrow e^+ + \pi^o + \nu + 358.3 \text{ MeV} \qquad (\sim 4\%)$$

The idea that the various empirical particles (θ, τ, etc.) are different manifestations of a single particle ran into difficulty. The difficulty was known as "tau-theta puzzle". Until 1956 it was assumed that parity was conserved in all nuclear and atomic processes. The two pion decay ($\theta^+ \rightarrow \pi^+ + \pi^o$) mode has even parity, but the parity of the three pion ($\tau^+ \rightarrow \pi^+ + \pi^+ + \pi^-$) mode is odd. There is no doubt that the theta and tau particles are both charged mesons with odd parity. Now difficulty is that the two identical particles end in states of different particles. This difficulty, known as "τ-θ *puzzle*", not yet been resolved but it led Lee and Yang in 1956 to consider the nature of the forces between particles and to suggest that weak interactions such as those involved in β-decay, parity would not be conserved.

An interesting situation arises in connection with the decay of the neutral K-mesons. Since K^o and $\overline{K^o}$ have different values of the hypercharge, + 1 and – 1 respectively, hence one cannot be converted into the other by the strong interaction. However, the K^o can be transformed into $\overline{K^o}$ by the *weak interaction*, For example :

$$K^o \rightarrow \pi^o + \pi^o \leftarrow \overline{K^o}$$

$$\rightarrow \pi^+ + \pi^- \leftarrow \overline{K}^o.$$

The transformation is possible only when CP symmetry invariance is applicable to the operation. Through the weak coupling between the K^o and its anti-particle the $\overline{K}^o$, it is possible to construct two neutral K-particles that are their own anti-particles. These states have wavefunctions

$$\psi\left(K_1^{\circ}\right) = \frac{\left[\psi(K^o) + \psi(\overline{K}^o)\right]}{\sqrt{2}} \quad ...(39)$$

$$\psi\left(K_1^{\circ}\right) = \frac{\left[\psi(K^o) - \psi(\overline{K}^o)\right]}{\sqrt{2}}. \quad ...(40)$$

Application of the CP operation to K_1°, i.e. converting K^o to $\overline{K}^o$ and vice versa, changes it into K_1°, i.e. K_1° is unchanged, but K_2° is changed into–K_2°, i.e. the sign of the wavefunction is changed. When the CP operation is applied to the decay products, it can be shown that decay of K_1° into $\pi^+ + \pi^-$ or $\pi^o + \pi^o$ is permitted. However, the K_2° particle, cannot decay by these two pion modes, although it can decay into three pions. The decay modes are :

$$K^o(K^o) \rightarrow \pi^o + \pi^o + 227.8 \text{ MeV } (31\%)$$
$$\rightarrow \pi^+ + \pi^- + 218.6 \text{ MeV } (69\%)$$

$$K^o\left(K_2^{\circ}\right) \rightarrow \pi^o + \pi^o + \pi^o + 92.8 \text{ MeV}(27.1\%)$$
$$\rightarrow \pi^+ + \pi^- + \pi^o + 83.6 \text{ MeV}(12.7\%)$$
$$\rightarrow \pi^{\pm} + e^{\mp} + \nu + 229.3 \text{ MeV}(36.6\%)$$
$$\rightarrow \pi^{\pm} + e^{\mp} + \nu + 252.5 \text{ MeV}(26.6\%)$$

The decay of a neutral K-meson into two pions is accompanied by a greater decrease in mass, and hence is a large energy release, then in the three pion decays. The K_1°-particles decay rapidly and have a short life than the K_2°. These states are also represented by K_S and K_L, where S and L indicate the shorter and longer lives respectively.

Mass of Kaons

Bristol and Brown analyzed forty seven τ^+-decay events and from all ranges and angles found Q = 75.04 ± 0.16 MeV. Using m_π = 139.58 ± 0.02 MeV, it was suggested that

$$m_{K+} = 493.78 \pm 0.17 \text{ MeV}.$$

Decay of neutral K°-meson in the mode $\pi^+ + \pi^-$ provides one means for measuring its mass. The estimated mass is 498.1 ± 0.4 MeV for K°-meson. Crawford found that close to the associated production threshold the energy balances of the reactions

$$\pi^- + p \rightarrow \Lambda + K^o$$
$$\rightarrow \Sigma^- + K^+$$

are sensitive to the mass values of the K-mesons. They found

$$m_{K^o} - m_{K^+} = 4.8 \pm 1.1 \text{ MeV}. \qquad ...(41)$$

Another approach is through the charge exchange reaction

$$K^- + p \rightarrow K^o + n.$$

Burnstein and Rubin were able to find mass difference

$$m_{K^o} - m_{K^-} = 3.90 \pm 0.25 \text{ MeV}. \qquad ...(42)$$

An earlier similar measurement by Rosenfeld et al. gave an estimate of 3.7 ± 0.7 MeV for this quantity. The mass of K°-meson thus calculated is

$$m_{K^o} = 497.7 \pm 0.3 \text{ MeV}.$$

The mass of the K^--menson has been determined by using the hyperon production reactions.

$$K^- + p \rightarrow \Sigma^+ + \pi^- + Q \qquad ...(43)$$
$$K^- + p \rightarrow \Sigma^- + \pi^+ + Q'. \qquad ...(44)$$

If the Q value of the eqn (43) is determined from the measured kinetic energies of the particles, the mass of the K^--meson can be found using the known mass of the Σ^+-hyperon.

$$m_{K^-} = 493.7 \pm 0.3 \text{ MeV (or } 965.8 \pm 1.5 \text{ } m_e).$$

K-meson (K^+, K^o) are strange particles and having strangeness quantum number +1.

Spin

The spin of the K-meson is zero, like of π-mesons. This can be obtained from any of the following arguments : (1) τ-decay does not involve a change of parity where as θ-decay does involve a parity change. This puzzle was resolved by Yang and Lee suggesting non-conservation of parity in weak interactions. The spectrum of the π^- in the 3π decay of K+-meson agrees with the theoretically predicted for a spiniess K-particle. (2) The observation of the neutral decay of the K_1° into π°-mesons is consistent only with an even spin of the neutral K-meson. (3) The longitudinal polarization of the μ^+-meson arising from the K^+-decay has been measured. It is in the same direction as the μ^+ from the decay of the π^+. This implies that the K^+ is also a spiniess particle.

Parity

The decay of the K^+ meson into three pions ($K^+ \to \pi^+ + \pi^+ + \pi^-$) can give us information about the parity of K^+. Two like mesons $\pi^+ \pi^+$ must have even parity $P = P_\pi P_\pi (-1)^L$, hence in this ($\pi^+ \pi^+$) system L must be even. The intrinsic parity of the pion is negative. Hence if l is the orbital angular momentum of the π^--meson, then the parity of the three pion system is $(-1)^3 (-1)^L (-1)^l = (-1)^{l+1}$. If spin of the τ^+ particle is zero, and the parity is conserved in the τ^+ decay, then its parity will be odd. In the two pion decay of θ^+ particle ($\theta^+ \to \pi^+ + \pi^+$), the total parity of the system is (–1) (–1) (–1)J. If parity is conserved in the θ-decay, the θ^+-particle parity is (–1)J. For a spiniess θ^+-particle, the parity is thus even. The even and odd parities of the two states of K^+-mesons were explained by Lee and Yang, who suggested that the τ^+ and θ^+ particles were identical particles and the parity conservation in K-meson decays was violated, as in β-decay.

The parity of the Kaon can only be determined in strong interactions. *All the Kaons ate found to have odd parity.*

Interaction with Matter

Charged K-mesons lose energy by ionization as they pass through matter and also have certain interaction with atomic nuclei. K^+-mesons undergo relatively little nuclear scattering or absorption and if brought to rest in matter undergo spontaneous decay as it would in the absence of matter. On the other hand, K^--mesons undergo quite strong nuclear scattering and absorption. If they come to rest in matter, they form K-mesic atoms and are absorbed, producing large stars. Neutral kaons do not appear to have very strong interaction with nuclei and do not lose energy by ionization. On account of their short lives they decay in flight.

For K^--energies less than 300 MeV, the reactions are

Elastic $\quad K^- + p \to K^- + p$

Charge exchange $K^- + p \to \overline{K^o} + n$

Inelastic $K^- + p \to \Sigma^- + \pi^+$

$\to \Sigma^+ + \pi^-$

$\to \Sigma^o + \pi^o$

$\to \Lambda^o + \pi^o$.

Both elastic and inelastic scattering are characterized by rapidly increasing cross-sections as the K^--energy decreases to zero. A very sharp K^--p scattering resonance has been found in elastic and inelastic cross-sections at about 135 MeV.

The K^--mesons interact with the nuclear neutrons as :

$$K^- + n \to \Lambda^o + \pi^-$$

$$\to \Sigma^o + p^-$$

K^--mesons produce hypernuclei, when interact with O^{16} and He^4. The resonance states of K and Λ (K* and Λ*) have been observed due to the interaction of K^- with protons. η^o-mesons are also produced when K^- interacts with protons.

CP Violation in Neutral K-Decay

We have seen earlier that the 2-pion mode is allowed by CP for the K_1, but forbidden for the K_2. The most sensitive test of the hypothesis of CP conservation is to look for the CP-violating decay $K_2 \to \pi^+\pi^-$. Christenson, Cronin, Fitch and Turlay in 1964 have performed an experiment and have observed the CP-violating decay. It was found in that experiment that K_2 particles decayed into $\pi^+\pi^-$ if with a branching ratio

$$\frac{\Gamma(K_2 \to \pi^+\pi^-)}{\Gamma(K_2 \to \text{all charged modes})} \cong (2.02 \pm (0.01) \times (10^{-3}). \quad ...(45)$$

This indicates that the violation is small. The charged particles are detected and their momenta measured, by the two spark chambers and magnet sandwiches. Most of the decays observed are the dominent K_2 modes (πμν and πeν). The π^--decays are recognized by computing $P_K = P_{\pi}^{+} + P_{\pi}^{-}$ and requiring that with in experimental errors resultant momentum vector P_K lies along the beam direction and the invariant mass of two charged particles directed be equal to the K^o mass. A second experiment has observed the same effect. A constructive interference between the free K_2-decay into two pions ($\pi^+\pi^-$) and the two pions ($\pi^+\pi^-$) from K_1's regenerated in baryllium has been also observed.

This CP violation suggests that the long lived and short lived components of neutral K's, known as K_L and K_S, are not the K_1 and K_2 combinations defined by equations (39) and (40). The above experiment suggests that

$$\psi(K_L) = \psi(K_2) + \in\psi(K_1), \quad \text{...(46)}$$

where $$|\in| = 2 \times 10^{-3}. \quad \text{...(47)}$$

The CP violation also implies that it is possible to make an absulute distinction between matter and anti-matter. With CP violation, $\overline{K^o}$ and K^o which are particle and anti-particle pairs can be distinguished by their decays into two pions. These are mixtures of K_S and K_L. Since K_S and K_L are not very different from K_1 and K_2. The relations are as shown by equations (39) and (40). Both K_S and K_L decay to two pions, so the time dependence of the 2-pion decay rate will involve are interference term, which has opposite sign for $\overline{K^o}$ and K^o.

η-MESON

Evidence of the existence of a neutral particle of mass ≈ 550 MeV disintegrating into 3π mesons was first reported by Pevsner and collaborators in 1961 and was named 11. It is a three-pion resonance or quasi-particles. It was observed in a deuteron bubble chamber exposed to a 1.23 BeV/e positive pion beam. The reaction is.

$$\pi^+ + d^+ \rightarrow p^+ + p^+ + \eta^o.$$

Its formation has been confirmed at Berkeley where this particle has been observed in the bombardment of protons by K^--mesons.

$$K^- + p^+ \rightarrow \Lambda^o + \eta^o.$$

Its chief modes of decay are

$\eta \rightarrow \gamma + \gamma$	(31.4%)	Electromagnetic interaction
$\eta \rightarrow \pi^o + \gamma + \gamma$	(20.5%)	Electromagnetic interaction
$\eta \rightarrow \pi^o + \pi^o + \pi^o$	(21%)	Weak interaction
$\rightarrow \pi^+ + \pi^o + \pi^-$	(22.4%)	Weak interaction

The η has a γ-width about 20 times larger than that of π^o. Considering that the γ-decay mode contributes roughly one half of the width, the mean life time η, obtained by comparison with that of π^o, should be 5×10^{-18} sec. This corresponds to the width of less than 1 keV.

Bastien et al. measured the invariant mass of the three pions in the reaction ($K^- + \pi \rightarrow \Lambda + \pi^+ + \pi^- + \pi^o$) and found the mass of η^o-meson as 549.9 ± 1.2 MeV. Delcourt et al. measured, with a quite different technique, the production of η-meson in the reaction $\gamma + p \rightarrow p + \eta$ just above threshold and estimated the mass 549.3 ± 2.9 MeV for η-mesons. The various published measurements are consistent. The best estimate derived is 548.7 ± 0.5 MeV.

As η, particle decays into two gamma rays or three pions, hence, it is necessary to assume the spin as 0, 2 or larger. Spin 2 or larger is excluded because it would result in marked anisotropies in the decay which are not observed. Thus it is better to assign J = 0 to the η-particle. Since the parity of two pions in a T = 0 state is necessarily even, hence the η-particle must have odd parity.

HYPERONS

Hyperons are unstable particles, having masses greater than that of the nucleon. These particles are a second class of strange particles. The name hyperon was taken from Greek word hyper, "meaning above".

Λ-hyperon

The first hyperon was found in cosmic rays in 1947 by Rochester and Butler in England. They observed the tracks of two charged particles-emerging from a point in the gas, of a cloud chamber. The upside down V-shaped figure could best be explained by assuming that some unstable neutral particle shooting down from the sky had decayed to form positive and negative charged particles. The charged particles seen in the decay were identified as a proton and a negative pion. This was the first evidence for the existence of the first and best known *hyperon. This neutral unstable particle was named as lambda and denoted by* Λ^o, *because of the characteristic appearance of its decay in a cloud chamber.* For its production in laboratories, Fowler, Shutt, Thorndike and Whittemore allowed a beam of 1.5 BeV π^--mesons to enter a diffusion cloud chamber containing hydrogen at a pressure of 18 atmospheres placed in a magnetic field pf 1100 gausses. The track of an incident pion can be seen at the end in the chamber. This end point is the vertex of a V, formed by the tracks of the negative pion and proton. To conserve energy and momentum in the production of Λ^o-hyperons, the necessity of the production of another neutral particle, probably K^o-meson, was also suggested. The reaction is

$$\pi^- + \pi \rightarrow L^o + K^o.$$

A-hyperon undergoes a following decay reactions :

$$\Lambda^o \rightarrow p + \pi^- \quad (67.7\%)$$
$$\rightarrow n^o + \pi^o \quad (31.6\%)$$

The mass of the Λ-hyperon has been obtained from the Q-value of the reaction $\Lambda^o \rightarrow p + \pi^-$. Both the pion and proton ranges, as well as the opening angles between the Λ-decay products were made and the Q-value so measured was 37.56 ± 0.13 MeV. Several other similar measurements have been made and the present best estimate of Q value is 37.60 ± 0.12 MeV. The corresponding mass

$$m_\Lambda = 1115.44 \pm 0.12 \text{ MeV}$$

The mean life of neutral lambda particle is 2.62×10^{-10} sec. The spin of Λ^o is 1/2. The magnetic moment of Λ^o, like the neutron is negative and equal to – (0.80 ± 0.07) nm.

The leptonic decay of the Λ^o-hyperon

$$\Lambda^o \rightarrow p + e^- + \overline{\nu}_e \quad (0.88 \times 10^{-50}\%)$$

has been observed to occur. Because of parity non-conservation in weak decay processes it is impossible to relate the parity of the Λ-hyperon to that of the πp system. By convention it is taken to have even parity. As Λ^o appears only in one charge state, neutral, hence it has isospin T = 0. The decay modes are permitted by the rule $\Delta S = 1$.

Σ-hyperons

It is the family of hyperons with the greatest number of members. The first of these to be observed was the positive sigma hyperon denoted by the symbol Σ^+. It has two predominant decay schemes and three weak decays.

$$\Sigma^+ \rightarrow p + \pi^o + 116.1 \text{ MeV} \quad (\sim 51\%)$$
$$\rightarrow n + \pi^+ + 110.3 \text{ MeV} \quad (\sim 49\%)$$
$$\rightarrow n + \pi^+ + \gamma \quad (\sim 4 \times 10^{-7}\%)$$
$$\rightarrow \Lambda^o + e^+ + \nu \quad (\sim 2 \times 10^{-7}\%)$$

$\rightarrow p + \gamma$ $(\sim 3 \times 10^{-5}\%)$

The mean life of Σ^+-hyperon is 0.81×10^{-10} sec. The corresponding negative particle denoted by Σ^- was also observed within few months of the discovery of the Σ^+. Its decay scheme is

$$\Sigma^- \rightarrow n + \pi^- + 118.5 \text{ MeV (Mean life time } 1.65 \times 10^{-10} \text{ sec).}$$

So other wards evidence for the existence of a neutral Σ^- hyperon was also presented. Its decay scheme is

$$\Sigma^o \rightarrow \Lambda^o + \gamma + 80 \text{ MeV (Mean life time} < 10^{-11} \text{ sec).}$$

The Σ-hyperon masses come from measurements carried out on the products of K^- and Σ^- capture by protons and on the disintegration products of the hyperons themselves. When such measurements are combined, they yield $m_{\Sigma+} = 1189.39 \pm 0.14$ MeV.

The π^- and Σ^+ ranges in the reaction

$$K^- + p \rightarrow \pi^- + \Sigma^+ + Q$$

indicate a momentum balance and yield a Q-value of 103.0 ± 0.3 MeV. Using this value of Q and the mass of the Σ^+-particle, the mass of K^- -meson is calculated. Then from relation $K^- + p \rightarrow \pi^+ + \Sigma^- + Q'$ we get the mass of Σ^- as 1197.6 ± 0.5 MeV. This is considerably larger than the mass of Σ^+.

Burnstein et al observed the sequence $\Sigma^- + p \rightarrow \Sigma^o + n$ followed by the decay mode $\Sigma^o \rightarrow \Lambda^o + e^+ + e^-$. These events yielded a Σ^o momentum of 60.3 ± 0.9 MeV/c and $m(\Sigma^-) - m(\Sigma^o) = 4.7 \pm 0.1$ MeV or $m(\Sigma^o) = 1192 \pm 0.2$ MeV.

The less probable decay modes of Z^- are :

$$\Sigma^- \rightarrow n + \mu^- + \overline{\nu_\mu} \qquad (6.6 \times 10^{-5}\%)$$

$$\rightarrow n + e^- + \overline{\nu_\mu} \qquad (1.4 \times 10^{-5}\%)$$

$$\rightarrow \Lambda^o + e^- \overline{\nu_e} \qquad (7.5 \times 10^{-7}\%)$$

Σ^+, Σ^- and Σ^o all have intrinsic parity even and intrinsic spin 1/2 thus are fermions. They possess a magnetic moment. The experiments on the Σ^+-hyperon magnetic moment suggested the value 3.28 ± 0.58 nm. Because of the three charge states, Σ-hyperons have T = 1. The decay modes show that $\Delta S = 1$. The Σ-parity can be related to that of the Λ by the analysis of the electromagnetic decay process $\Sigma \rightarrow \Lambda\, e^+e^-$

Σ^--hyperons undergo interactions with nucleons

$$\Sigma^- + p \rightarrow \Lambda^o + n.$$

Ksi-hyperons

Ksi-hyperons known thus far are the Ξ^- and Ξ^o. Present theory of elementary particles predicts that there should be no positively charged Ξ particle. The Ξ's were originally called *cascade particles*, because a Ξ particle decays to produce π-meson and a Λ^o, which in turn decays into a nucleon and another π-meson. The known modes of decays of these particles can be represented by the eqautions

$$\Xi^- \rightarrow \Lambda^o + \pi^- + 66.3 \text{ MeV}$$

$$\Xi^o \rightarrow \Lambda^o + \pi^o + 70 \text{ MeV}$$

The decay processes have been used most often for mass measurements, but a good check was made by Schneider who measured the mass of Ξ^- hyperon in the production reaction ($K^- + p \rightarrow K^+ + \Xi^-$) Using beam energy of 35 MeV, he yielded $m(\Xi^-) = 1321.1 \pm 0.6$ MeV. Rosenfeld et al suggested $m(\Xi^-) = 1320.8 \pm 0.2$ MeV and $m(\Xi^o) = 1314.3 \pm 10$ MeV.

The Ξ^--hyperon has mean lifetime of 1.3×10^{-10} sec. and the Ξ^o-hyperon has its mean lifetime as of the order of 10^{-10} sec.

These hyperons are produced by the interaction of π^- and K-mesons of sufficiently high energy with nucleons. These mesons are absorbed by the nucleus and undergo one of the following reactions:

$$
\begin{aligned}
K^- + p^+ &\rightarrow \Lambda^o + \pi^o \\
&\rightarrow \Lambda^o + \pi^+ + \pi^- \\
&\rightarrow \Lambda^o + \pi^o + \pi^o \\
&\rightarrow \Sigma^+ + \pi^- \\
&\rightarrow \Sigma^- + \pi^+ \\
&\rightarrow \Sigma^o + \pi^o \\
&\rightarrow \Sigma^- + K^+ \\
&\rightarrow \Xi^o + K^o \\
K^- + n &\rightarrow \Lambda^o + \pi^- \\
&\rightarrow \Lambda^o + \pi^- + \pi^o \\
&\rightarrow \Sigma^- + \pi^o \\
&\rightarrow \Sigma^o + \pi^- \\
&\rightarrow \Xi^- + K^o \\
&\rightarrow \Xi^o + K^- \\
\pi^- + p &\rightarrow \Lambda^o + K^o
\end{aligned}
$$

$$\rightarrow \Sigma^o + K^o$$

$$\rightarrow \Sigma^- + K^+.$$

In all of the above reactions considerable energy is released axid the nucleus in which the absorption occurs, therefore, breaks up into several pieces. If one piece is a hyperfragment, the Λ^o, it must eventually decay into a proton and a pion, each of which may either interact with the residual nucleus or escape with its full energy.

Ξ^- and Ξ^o particles have spin 1/2, and arc thus fermions. They possess a magnetic moment. Ξ^- and Ξ^o have been observed with T = 1/2 and thus $T_3 = -1/2$ and + 1/2. Their production and decay reactions show that the strangeness quantum number S=–2 for these two and $\Delta\Sigma = 0$.

Ω–hyperon

Before the discovery of Ω^--hyperon, all known particles had hypercharge values of + 1, 0, or – 1. Gell-Mann predicted the existence of such a particle, with a negative charge, as required to give it a hypercharge of –2. Furthermore, he suggested that it might be produced, as it actually was, by the interaction of a K^--meson at energies greater than 3.5 GeV with a proton. The expected strong interaction would be

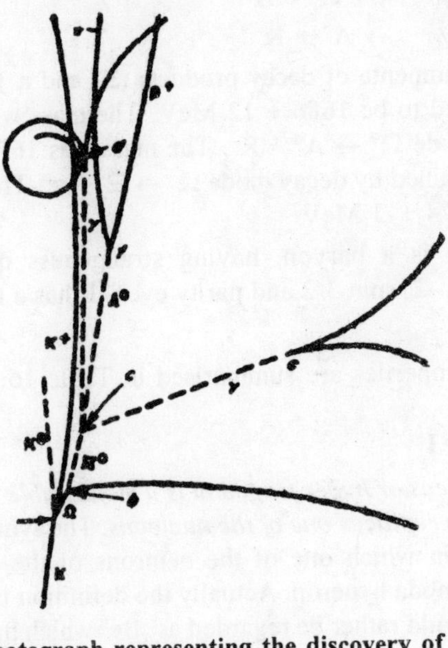

Fig. 12 : Photograph representing the discovery of Ω^--hyperon.

$$K^- + p^+ \rightarrow \Omega^- + K^+ + K^o$$

Hypercharge − 1 + 1 − 2 + 1 + 1

Isospin 1/2 1/2 0 1/2 1/2

The experimental designed to produce and detect the Ω^--hyperon was carried out at Brookhaven National Laboratory in 1964. In this experiment short bursts of 5 GeV K^--mesons were allowed to enter a liquid hydrogen bubble chamber. The photograph obtained is shown in Fig. 11 and can be interpreted as : *A K^--meson enters from the bottom and interacts with a proton at the point a to yield a K^o-meson, a K^+-meson and the Ω^--hyperon.* The positive kaon is deflected to the left by the magnetic field whereas the Ω^--hyperon is deflected to the right. The Λ^o-hyper on decays at point b, after travelling a short distance during its lifetime, into Λ^o hyperon and π^--meson. Ξ^o-hyperon decays at point c, into Λ^o and π^o, but latter forms two gamma ray photons immediately. These photons then produce electron-position pairs at the points d and e. The Λ^o-hydrogen decays at point f into a proton and a π^--meson.

Gell-Mann noted that the Ω^--particle should decay by the weak interaction, which it does. Its decay modes are

$$\Omega^- \rightarrow \Xi^o + \pi^-$$
$$\rightarrow \Xi^- + \Lambda^o$$
$$\rightarrow \Lambda^o + K^-$$

From the momenta of decay products (Ξ^o and π^-), the mass of the Ω^- was measured to be 1686 ± 12 MeV. The mass was also measured by the decay mode $\Omega^- \rightarrow \Lambda^o + K^-$. The mass was 1674 ± 3 MeV. The mass was determined by decay mode $\Omega^- \rightarrow \Xi^- + \pi^o$. The averaged value used now is 1674 ± 3 MeV.

Ω^--hyperon is a baryon, having strangeness quantum number −3. hypercharge −2, spin 3/2 and parity even. It has a mean life time of 1.3×10^{-10} sec.

All these properties are summarised in Table 16.3.

HYPERNUCLEI

A hypernucleus or hyper fragment is a metatstable nucleus in which a bound hyperon replaces one of the nucleons. The symbol ${}_{\Lambda}B^{10}$ denotes a hypernucleus in which one of the neutrons of the ${}_5Be^{10}$ particle is replaced by a lambda hyperon. Actually the definition is misleading, the hypernucleus should rather be regarded as ${}_5Be^9$ which has captured a Λ^o-particle. The reason for this is that the Λ^o can occupy the same quantum state as a neutron already in the nucleus without violating the Pauli exclusion principle, since it is a different kind of particle. We shall restrict the discussion to Λ-hypernuclei. Because of the fast transformations strangeness conserving Σ or Ξ-hypernuclei have very short lifetimes and interact with nucleons as

$$\Sigma^+ + n \rightarrow \Lambda^o + p$$
$$\Sigma^- + p \rightarrow \Lambda^o + n$$
$$\Xi^- + p \rightarrow \Lambda^o + \Lambda^o$$
$$\Xi^o + n \rightarrow \Lambda^o + \Lambda^o.$$

The Polish physicists Danysz and Pniewski in 1952 deduced that a track observed in a nuclear emulsion block which had been exposed at a high altitude to cosmic radiation was due to hypernucleus of Beryllium. Hypernuclei of all the light elements from hydrogen (${}_{\Lambda}H^3$) to nitrogen (${}_{\Lambda}N^{14}$) have been reported. In addition, from a star produced in a photographic film bombarded by K^--mesons, evidence of a hypernucleus containing two Λ^o-hyperons was reported in 1963 at the CERN laboratory. The species formed was considered to be either ${}_{\Lambda\Lambda}Be^{10}$ or ${}_{\Lambda\Lambda}Be^{11}$. In 1966, a similar hypernucleus ${}_{\Lambda\Lambda}He^6$ was identified in U.S.A.

It is very difficult to obtain accurate determination of the average life-times of hypernuclei, but the average life-times are approximately slightly less than that of a free Λ^o-hyperon. There are usually several competing modes of decay of a given species of hypernuclei. The ${}_{\Lambda}He^4$ hypernucleus has the following decay modes:

$$\begin{aligned} {}_{\Lambda}He^4 &\rightarrow \pi^o + \text{nucleons} \\ &\rightarrow \pi^- + \pi + {}_2He^3 \\ &\rightarrow \pi + n + {}_1H^2 \\ &\rightarrow {}_1H^2 + {}_1H^2. \end{aligned}$$

The first two decay modes are called mesonic decays, in which hyperon may be regarded as undergoing its normal decay into $p + n^-$ or $n + \pi^o$. The nucleon thus produced may be expelled or it may remain in the product nucleus. Other two modes are called non-mesonic decays. The non-mesonic decay is equivalent to the interaction of a Λ^o-particle with proton or a neutron, i.e.

$$\Lambda^o + p \rightarrow (\pi^- + p) + p \rightarrow n + p$$
$$\Lambda^o + n \rightarrow (\pi^o + n) + n \rightarrow n + n.$$

Since a Λ^o-hyperon does not normally interact with a nucleon, the interaction with a nucleon is stimulated by the remainder of the nucleus and the non-mesonic decay (dominant for hyper-nuclei with $Z > 3$ or 4) is referred to as stimulated decay.

The binding energy of the Λ^o-particle in the hypernuclei can be calculated from the observed kinetic energies of the decay particles, which can be derived from the lengths of the tracks in a photographic emulsion. In non-mesonic decay, one or more neutrons are generally emitted, they do not produce any track in the emulsion and the kinetic energies cannot be determined. Hence, the binding energy of Λ^o-particle is obtained from mesonic decays. The values have been found to increase fairly regularly with the atomic number of the hypernucleus. The value is 0.2 MeV in ${}_{\Lambda}He^3$ and about 12 MeV in ${}_{\Lambda}C^{14}$ and ${}_{\Lambda}N^{14}$. In this way it is clear that the existence of ${}_{\Lambda}H^2$ is not possible because in it the binding energy would be still smaller and in the heavier hypernuclei, Λ^o-hyperon is more strongly bound than a nucleon. Similar to with nucleon–nucleon force, the hyperon-nucleon force appears to be charge independent and dependent on the spin.

SOLVED EXAMPLES

Example 1:

Calculate the energy of the neutron produced when a slow negative pion is captured by a proton. Should neutron be treated relativistically?

Solution:

$$\pi^- + p^+ \rightarrow n^o + \gamma + Q.$$

$$139 + 938 \rightarrow 939 + h\nu + Q,$$

or $$Q = 138 \text{ MeV} = E_\gamma + E_n.$$

From conservation of momentum $m_n v_n = E_\gamma/c$

and $$\frac{E_n}{E_\gamma} = \frac{1}{2}\frac{m_n v_n^2}{c m_n v_n} = \frac{1}{2}\frac{v_n}{c} = \frac{1}{2}\frac{E_\gamma}{m_n c^2} = \frac{E_\gamma}{1878}$$

$$\frac{E_n}{E_\gamma + E_n} = \frac{E_\gamma}{1878 + E_\gamma} \quad \therefore \quad \frac{E_n}{138} = \frac{E_\gamma}{1878 + E_\gamma}$$

and $$\frac{E_n}{138} = \frac{138 - E_n}{1878 + 138 - E_n} = \frac{138 - E_n}{2016 - E_n}$$

or $$E_n = 9 \text{ MeV}.$$

Using relativistic relations

$$m_n v_n = \frac{m_{0n} v_n}{(1-\beta^2)^{1/2}} = \frac{E_\gamma}{c}$$

and $E_n = m_{0n}c^2 \, [(1 - b^2)^{-1/2} - 1]$

We get $\quad E_n = 8.8$ MeV.

Example 2:

Discuss the following reactions using the conservation of isotopic spin : (i) $p + p \rightarrow p + p + \pi^o$, (ii) $\pi^o \rightarrow \gamma + \gamma$, (iii) $\Lambda^o \rightarrow \pi^- + p^+$, and (iv) $\pi^+ + p \rightarrow \pi^+ + p$.

Solution:

(i) $\quad p + p \rightarrow p + p + \pi^o$

$$T = \quad \frac{1}{2} \quad \frac{1}{2} \quad \frac{1}{2} \quad \frac{1}{2} \quad 1$$

$$T_3 = + \quad \frac{1}{2} + \frac{1}{2} + \frac{1}{2} + \frac{1}{2} \quad 0$$

We know that in all strong and electromagnetic reactions the third component of isotopic spin is conserved. As $Q = e\,(T_3 + B/2 + S/2)$ must be conserved, hence change of T_3 will be there when strangeness changes, which happens in the weak interactions. Secondly, in combining isotopic spins, we use the same rules as for ordinary spin. Hence vector sum of T on the right band side may be 2, 1, or 0. Again the total $T_3 = 1$, ruling out T = 0. The conservation law can be satisfied with the final T = 1.

(ii) $\pi^o \rightarrow \gamma + \gamma$

$T = 1 \qquad$ no T

$T_3 = 0$

In this decay, T_3 is conserved hence the process does not involve the weak interaction. As the total isotopic spin changes, the decay occurs through the electro-magnetic forces which do not conserve isotopic spin.

(iii) $\quad \Lambda^o \rightarrow \pi^- + p^+$

$$T = 0 \quad 1 \quad \frac{1}{2}$$

$$T_3 = 0 \quad -1 \quad +\frac{1}{2}$$

The decay of the Λ^o is a weak interaction which can bypass several conservative laws. The π^- is emitted in one direction from the plane defined by the Λ^o and the bombarding particle which produced it. Thus parity is not conserved. As T_3 does not conserve, hence the conservation of charge gives no conservation of strangeness.

(iv) $\pi^+ + \quad p \rightarrow \quad \pi^+ + \quad p$

$$T = 1 \quad \frac{1}{2} \qquad 1 \qquad \frac{1}{2}$$

$$T_3 = +1 \quad +\frac{1}{2} \qquad +1 \qquad +\frac{1}{2}$$

In this strong reaction both T_3 and T must be conserved. In the case of the π^+ rection only one isotopic spin state is possible.

lowing processes in terms of the type of interaction.

$\pi^- + p \rightarrow \Lambda^o + K^o$; $\pi^- \rightarrow \pi^o + n$; $p + \gamma \rightarrow p + \pi^o$; $\Sigma^o \rightarrow \Lambda^o + \gamma$; $\pi^o \rightarrow \gamma + \gamma$; $K^o \rightarrow \pi^+ + \pi^-$; $\Lambda^o \rightarrow p + \pi^-$; $\Xi^- \rightarrow \Lambda^o + \pi^-$; *and* $\Lambda^o \rightarrow p + e^- + \bar{\nu}$.

Solution:

In the reactions $\pi^- + p \rightarrow \Lambda^o + K^o$ and $\pi^- + p \rightarrow \pi^o + n$, the interaction is a short-range force between hadrons {π, p, n, Λ and K) corresponding to one pion-exchange. These reactions obey selection rules

$$\Delta B = 0, \Delta Q = 0, \Delta Y = 0, \Delta \pi = 0, \Delta T = 0, \Delta T_3 = 0$$

hence the interaction is *strong*.

In the reactions $p + \gamma \rightarrow \pi^o + p$, $\Sigma^o \rightarrow \Lambda^o + \gamma$ and $\pi^o \rightarrow \gamma + \gamma$, the interaction is electromagnetic, as the electromagnetic force arises due to the mechanical effect of the emission and absorption of virtual photons.

These reactions obey all above mentioned selection rules except the rule for conservation of isospin T, hence the interaction in these processes is *electromagnetic*.

In the reactions $K^o \rightarrow \pi^+ + \pi^-$, $\Lambda^o \rightarrow p + \pi^-$, $\Xi^- \rightarrow \Lambda^o + \pi^-$ and $\Lambda^o \rightarrow p + e^- + \bar{\nu}$, the interaction is weak, as the weak interaction is due to the leptonic decay of strange and non-strange particles and due to the non-leptonic decay of strange particles. These reactions do not conserve parity, strangeness, isospin and the third component of isospin, hence the interaction in these processes is weak.

EXERCISES

1. Complete the following reaction formula, taking account of conservation of charge, baryon number and strangeness (More than one answer is possible and more than one particle is necessary).

   $p + p \rightarrow \Xi^o +$

2. Which of the following reactions illustrate associated production of strange particle?

   (i) $p + n \rightarrow p + \Sigma^- + K^+$

   (ii) $p + \gamma \rightarrow p + \mu^+ + \mu^-$

   (iii) $p + p \rightarrow n + \Sigma^+ + K^o + \pi^+$.

3. Which of the following reactions and decay processes strangeness and which interaction is responsible for each process?

   (a) $\pi^o \rightarrow \gamma + \gamma$

   (b) $\mu^- + p \rightarrow n + \nu_\mu$

   (c) $\Lambda^o \rightarrow p + \pi^-$

   (d) $\Lambda^o + p \rightarrow S^+ + n$

   (e) $p + n \rightarrow p + p + \pi^-$.

4

Nuclear Emulsions and Models

INTRODUCTION

The discovery of radioactivity by Becquerel (1896) can be explained as charged particles pass through a photographic emulsion of a photographic plate they interact with the tiny silver halide crystals producing a line of latent images along their path. In 1911, M Reinganum in Germany recognized that individual α-particles traversing a photographic emulsion produced separate tracks in a photographic emulsion as a series or closely spaced black specks of silver. Because of the relatively high density and stopping power of the photographic emulsion, the tracks produced by charged particles were very short compared with those obtained in a cloud chamber but do not differ greatly from bubble tracks. Miss Blau and Wambacher, in 1937, discovered 'Star' due to the interaction of high energy cosmic rays with the atomic nucleus in the emulsion.

The optical photographic emulsions is not suitable for quantitative work with nuclear radiations. The sensitivity is low and the tracks due to charged particles have non-clear range because the developed crystal grains are large and widely spaced. The composition of the emulsion was changed so as to make it more suitable for study of various ionizing particles, such as alpha particles, protons, mesons and even electrons The nuclear emulsions differ from the optical emulsions in that they have considerably higher silver halide content and smaller grain size In nuclear emulsion the thickness is greater than that of optical emulsion. The smaller the grain size, the more sensitive the emulsion to ionizing particles. Thus different commercially available emulsions, differing chiefly in grain size, can be used to discriminate between different particles.

Table 1 : Composition in kg/m³ of Dry Nuclear Emulsions

↓ Element→ Type	H	C	N	O	S	Br	Ag	I
Ilford	49	300	73	200	11	1465	2025	57
Kodak Ltd.,	38	270	80	160	–	1440	1970	36
Eastman Kodak	43	340	110	170	–	1220	1700	54

The least sensitive emulsions will show only fission fragment tracks, whereas the most sensitive ones reveal the tracks of singly charged minimum-ionizing particles. A series of nuclear emulsions has been developed for detecting particles of different specific ionization. D-l emulsion is the suitable type for detecting fission fragments and alpha particles. The C-2 Ilford emulsion is a useful general purpose emulsion, suitable for recoiling mesons, protons and heavy nuclei. The Ilford G-5 is used to record high energy electrons. The tracks produced in an emulsion are very short. They can be magnified and photographed. High quality microscopes are required for sinning the developed plates.

Table 2 : Nuclear and Optical Emulsions

Property	Optical Emulsion	Nuclear Emulsion
AgBr : Gelatin (mass)	4 : 53	80 : 20
AgBr : Gelatin (volume)	15 : 85	45 : 55
Grain size (micron)	1—3.5	0.1—0.6
Thickness (")	2—3	25—2000
Sensitivity to light	Very high	Poor
Response to α-particles	Dense blackening	Individual tracks
" " β "	Moderate "	Faint fog
" " γ "	Faint "	Almost none

Table 3 : Characteristics of Nuclear Emulsions

Ilford Emulsion type	D1	E-1(K1)	C_2(K2)	B-2	G-5
Mean grain diameter (mm)	0.12	0.14	0.10	0.21	0.18
Highest detectable velocity β	0.2	0.31	0.46	all	
Highest detectable energy of protons in MeV		20	50	120	all

The nuclear emulsion with their high silver halide concentration shrink during the fixation process unfortunately. A knowledge of this shrinkage factor is necessary when we want to determine the angle of dip of a track in the emulsion.

Fast neutrons can be detected by the knock on protons in unloaded emulsions. Slow neutrons can be detected by using emulsions impregnated with lithium, boron or uranium. Gamma rays of energy above about 3 MeV can be detected through the track of the protons, released in photo-disintegration reaction, in emulsions loaded with deuterium compounds.

Under favourable conditions it is possible to determine the charge, mass and incident K.E. of a charged particle brought to rest in a nuclear emulsion. The range of the particle, the grain density and its variation along the track and the small scattering angle due to collisions of the particle with nuclei in the emulsion are all relevant to these determinations. We shall confine ourselves here to discussion of the main relationships which have been established by experiments.

The number of grains deposited per micron dn/dx of track is closely related to the energy loss per unit of track dT/dx keV/micron. These two quantities are related by the empirical relation

$$\frac{dn}{dx} = a\left\{1 - e^{bz(dT/dx^{1/2} - c^{1/2}}\right\}, \qquad ...(1)$$

where a, b and c are empirical constants and z is the integer characterizing the charge of the ionizing particle.

If m (the mass in units of proton mass) and z of a particle are known, its kinetic energy T (MeV) may be obtained from its total range R (microns) by an equation

$$T = Kz^{2n}\, m^{1-n}\, R^{n}, \qquad ...(2)$$

where K and n are empirical constants.

The total number of silver grains deposited and the total particle range are related by an equation

$$N = K'\, z^{2n'}\, m^{1-n'}\, R^{n'}, \qquad ...(3)$$

where K' and n' are another constants.

Fig. 1 gives the range of protons, deuterons, α-particles, pions and muons in Eastman Kodak NTA emulsions as a function of particle energy.

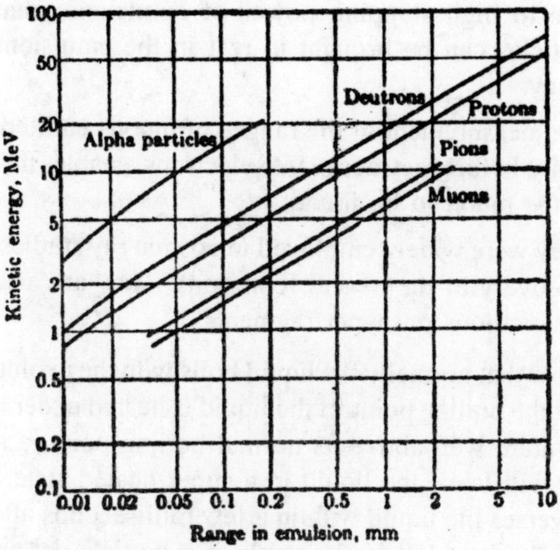

Fig. 1 : Range V/s energy for various particles in nuclear emulsions.

As compared with track chambers, like bubble and cloud chambers, photographic emulsions have following *drawback.* It is very difficult to determine the sign of the charge and the momentum of a particle from observations of the curvature of the path in a strong magnetic field. The actual path length in the emulsion is small compared to that in a cloud chamber and the multiple Coulomb scattering in the emulsion stimulates the curvature. By the use of very strong magnetic fields, the deflection of long (but not short) tracks in emulsions has been observed. This gives the sign of the charge, but the large amount of scattering makes the deflected tracks useless for the momentum calculations.

Some of the *advantages* of the nuclear emulsion detector are:

1. It produces a visible track following the complete trajectory of a charged particle. Particles of different ionizing powers produce tracks which appear quite different.
2. The emulsion is continuously sensitive and is consequently always available to record an event.
3. The energy of product particles can be determined from their range in the emulsion and the differential scattering cross section from counts of the number of tracks formed in each emulsion.
4. The emulsion is relatively light and cheap. This makes it better for high altitude cosmic ray experiments.

5. Due to high stopping power of emulsions, many short lived particles can be brought to rest in the emulsions, before they decay.
6. The measurement of the ranges of muons emitted from stopped pions, before they decay into electrons, enables the mean lifetime of the muon to be found.
7. They were widely employed in cosmic ray studies and led to the discovery of the π- and K-mesons. They are also employed in observations on hyperfragments.

We know that normally the liquid boils with the evolution of bubbles of vapour at the boiling point. If the liquid is heated under a high pressure to a temperature well above its normal boiling point, a sudden release of pressure will leave the liquid in a super heated state. If an ionizing particle traverses the liquid with in a few milliseconds after the pressure is released, the ions left in the track of a particle act as condensation centres for the formation of vapour bubbles. The vapour bubbles grow at a rapid rate and attain a visible size in a time of the order of 10 to 100μ sec. Upon this principle chamber operates. The bubbles are like the droplets in a cloud chamber, visible under strong illumination. If nuclear reactions take place in the liquid of the bubble chamber, the sets of tracks diverge from the collision centre as in the cloud chamber. The tracks are photographed against a dark background.

BUBBLE CHAMBER

To detect very energetic particles the Cloud Chamber is not very suitable since the interaction of energetic particles cannot be completely observed in the chamber. To overcome this disadvantage it is necessary to use a chamber filled with a substance having a much greater stopping power. This has led to the development of the Bubble Chamber. In 1952, D.A. Glaser, at the University of Michigan, designed and made the first bubble chamber and this has now become one of the largest and at the same time one of the costiliest type of particle detector in use.

The time sequence of events for a typical chamber is shown in Fig. 2. At the instant A, the high pressure P_1 (400 units) applied to the bubble chamber liquid is suddenly reduced to the lower pressure P_1 (100 units). The pressure starts to rise due to the bubble formation. The liquid is recompressed at C and the chamber returns to its initial position at A'. The time interval between A and A' gives the cycling rate of the chamber. The expansion is timed so that the particle enters the chamber at X and the bubbles are illuminated and photographed at the moment Y.

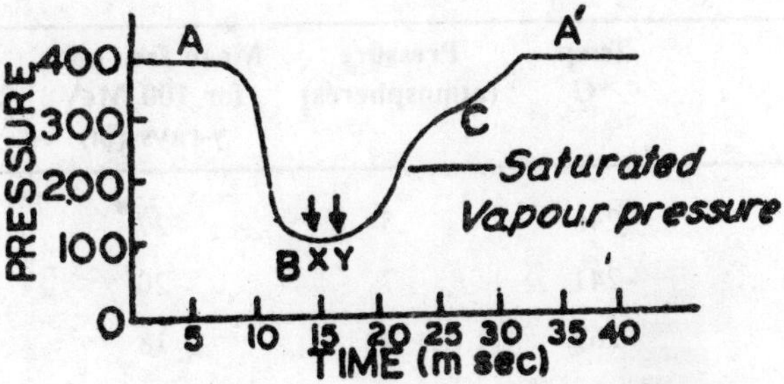

Fig. 2 : Time sequence of events in bubble chamber.

The liquids suitable as the working substance of a chamber are transparent and having a low surface tension and a high vapour pressure. The medium in first bubble chamber was ordinary ether, *i.e.,* diethyl ether, but liquid hydrogen, deuterium, helium, propane, xenon and other liquids have been utilized successfully for different purposes. In some experiments it is desired *to study the interaction of particles with protons*, the liquid in the chamber is then hydrogen. Propane (C_3H_8) is commonly used in the bubble chamber to avoid the need for low temperature (–246°C) operation with liquid hydrogen. It has almost 1.3 times as many hydrogen nuclei (*i.e.,* protons) per unit volume, but the carbon nuclei present are sometimes a drawback. For *observations on neutron interactions* liquid deuterium with a neutron fairly loosely bound to a proton if employed.

Liquid helium chambers operating at 4°K and 1 atm. have been used for studying high energy interactions with the nucleons in the He^4 nucleus. A liquid containing heavy atoms is favoured where stopping power is the main objective, *e.g., in studying the interactions of neutral particles and gamma rays.*

Examples are the freon type compound, trifluoro-bromomethane (CF_3Br) and liquid xenon. The former can be used just above room temperature (30°C, 18 atm.) however there are three distinct complex nuclei present in this liquid. Liquid xenon operating at 14°C is ideal for studying interactions with high energy γ-quanta and neutral pions. For satisfactory bubble tracks a small amount of ethylene is to be added. The operating conditions for bubble chambers with different fluids are given in the following Table 4.

Table 4 : Operating conditions for Bubble Chamber fluids

Fluid	Temp. °C	Pressure (atmospheres)	Mean free path for 100 MeV γ-rays (m)	Density (Kg m^{-3})
Hydrogen	–246	5	27	60
Deuterium	–241	7	20	130
Helium	–269	1	18	130
Propane	58	21	2.2	430
Pentane	157	23	—	500
Xenon	–20	26	0.07	2300

Liquid hydrogen is an ideal substance for studying high energy interactions with protons, as it is a pure proton target having very small Coulomb scattering. L W. Alvarez at the Lawrence Radiation Laboratory, California University, thought that if the chamber was sufficiently large and the pressure of the compressed liquid was decreased rapidly, good tracks could be obtained in the interior in spite of the fact that bubbles were formed at the walls. This conjecture proved to be correct and a chamber six feet long and of cross-section 1½ feet, containing 520 litres of liquid hydrogen, was first operated in 1959 successfully. A schematic diagram of a hydrogen bubble chamber, operating at a temperature of 27°K and developed by Alvarez and his co-workers is illustrated in Fig. 3.

The vessel was made of stainless steel with glass ports at the top for the viewing cameras. A box of thick glass walls was filled with liquid hydrogen and connected to the expansion pressure system. To maintain the chamber at constant temperature it was surrounded by liquid nitrogen and liquid hydrogen shields. From the side window W the high energy particles were allowed to enter the chamber. Liquid hydrogen was kept under pressure first but when a particle has passed through it the pressure was released so that it was super heated at a lower pressure. A sudden release of pressure from the expansion valve we followed by light flash and camera took the steroscopic view of the chamber.

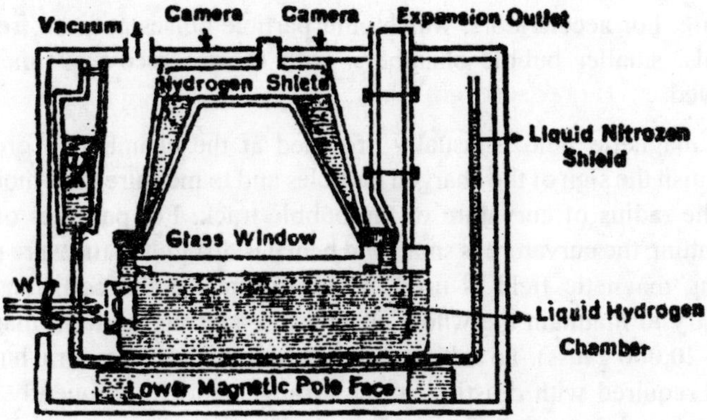

Fig. 3 : Schematic diagram of Alvarez bubble chamber.

This was the first of the so-called dirty bubble chambers. Subsequently, 80 inch hydrogen bubble chamber was built at the Brookhaven National Laboratory and similar one is in use at CERN (*Conseil Europeen Pourala Recherche Nucleaire*). Even larger bubble chambers, upto 14 feet in length, are being designed. Similar to the original cloud chamber bubble chamber is not continuously sensitive.

(i) A dirty bubble chamber is sensitive for only a few milliseconds after the pressure reaches its minimum.

(ii) The relatively slow expansion of the liquid does not permit the process to be initiated by an ionizing particle continuously.

(iii) The bubbles formed spontaneously at discontinuities and they interfere with the photography of the particle tracks.

(iv) Once the liquid has been expanded, it takes some time to recover sufficiently for another set of tracks to be recorded.

These drawbacks are overcome by using bubble chambers in connection with high energy particle accelerators, which function in a pulsed manner. The chamber can be made sensitive each time a pulse of particles emerges from the accelerator, and the large density and large volume of the chamber permit appreciable numbers of interesting collision processes even when the incident particles have energies of thousands of MeV. The timing is such that the particle beam from the accelerator enters the bubble chamber just prior to the pressure minimum. A few milliseconds later, powerful lights flash on and the tracks are recorded on a photographic film which moves forward automatically after each exposure. For accelerators, which emit particle pulses at more frequent intervals, smaller bubble chambers, with shorter recovery times, are employed.

A magnetic field is usually provided at the chamber in order to distinguish the sign of the charged particles and to measure their momenta from the radius of curvature of the bubble track. For particles of high momentum, the curvature is small and hence is difficult to measure unless a strong magnetic field is used. Thus, to analyse the collisions it is necessary to maintain the whole chamber in a strong uniform magnetic field (~20,000 gauss). To achieve this a magnet weighing several hundred tons is required with drastic cooling arrangement. Consequently, there has been much interest in the possibility of constructing superconducting magnets for use of bubble chambers. The problem in the construction of such superconducting magnets of appreciable size has been the difficulty of producing cable with satisfactory physical and mechanical properties at a reasonable cost. In recent years, certain metallic compounds (alloys) of niobium, *e.g.*, with tin, titanium, or zirconium, have been developed for fabrication into super conducting cables. The first successful operation of a bubble chamber, 10 inches in diameter, with a super conducting magnet was achieved at the Argonne National Laboratory in 1966.

The analysis is laborious and is greatly assisted by computing machines. A simplified account of the methods used is given below. The first step in the analysis, involving the selection of events of significance to the experimenter, is known as *scanning*. The magnified image of the film is inspected and the locations of interesting events are recorded. The next stage in track analysis, known as *tracking*, involves various degrees of automation. The tracking operation is greatly expedited by utilizing a flying spot digitizer. A small spot of light is moved rapidly in this system either mechanically or electronically, across and down the image of the bubble chamber event. Each time the moving spot of light crosses a track, co-ordinates of the point are expressed in digital form. The complete co-ordinates of the event of interest are next treated by a computer to yield what is called the *geometry* of the event. For the next stage, known as *kinematic analysis*, it is required to know the masses of the particles that produce the observed tracks. The kinematic analysis, made by the computer at each vertex of an event is based on the conservation laws of momentum and of kinetic energy.

Nevertheless, cloud and bubble chambers both are important detectors of radiation from a historical point of view since they have played such an important part in the study of fundamental particles and their interaction. They are used mainly in Nuclear Physics research. It has been estimated that a comparatively small bubble chamber is equivalent in stopping power to a cloud chamber some 140 ft. long.

SPARK CHAMBER

The spark chamber utilizes an incipient electrical discharge in a gas, like that used in a Geiger counter, to give track geometry information, like that provided by a bubble chamber. This has found increasing use in the field of high energy physics. The basic principle of operation is as follows:

A high voltage is maintained between two plates spaced in a vessel containing a gas but the electric field is not quite enough to permit the passage of a spark. Now if an ionizing particle enters the gas space, a spark will pass and it will tend to follow the path of the ion pairs produced by the ionizing particle. The passage of a spark through a gas that had been ionized by nuclear radiation was used by H. Greinacher, in 1935, for the detection of α-particles. F. Bella and C. Franzinetti in 1953, in Italy, confirmed the localization of the spark discharge and the first photographs were given in 1955 by P.G. Henning in Germany. The modern spark chamber stems from the work described by the British physicists T. E. Cranshaw and J. F. de Beer in 1957. In 1959, S. Fukui and S. Miyamoto in Japan found that with neon several particle tracks could be observed simultaneously in the spark chamber and with air only one particle track could be formed at a time. Spark chambers were used by Sree Kantan and coworkers at T. I. F. R. Bombay in 1960, to trigger high energy cosmic-ray events.

In its simplest form spark chamber consists of a series of large thin parallel metal plates (aluminium), several square feet in area, spaced about 1 cm. apart in a chamber filled with a neon gas at atmospheric pressure. All the plates are isolated from each other. The first, third, fifth, etc., plates are grounded and the second, fourth, sixth, etc., plates are connected to a high voltage direct current pulse generator, which gives them a high potential in short bursts of the order of a microsecond each. This potential is just enough to cause sparks to occur between the plates in such regions as are ionized by a particle entering the chamber. The chamber is sensitive for about half microsecond preceding the application of the voltage.

Any charged particle passing through the chamber during this period produces along its path a visible and audible spark discharge, which can be located by photographing or by other methods. The photograph of all the individual spark reveals the sections of the trajectory of the particle between the plates. Ordinarily two cameras at right angles to one another are used to photograph.

In order to prepare the chamber for recording the path of another particle, any ion-pairs present between the plates must be removed, by applying a clearing field of about 200 volts per cm. continuously in the direction opposite to that of the high voltage spark field. Due to this field a spark can form only if the high voltage is attained within about a microsecond after the entry of an ionizing particle. In practice this delay time is about 0.2 microsecond. The recovery time depends on the gas, the design of the chamber and the energy of the spark. It is usually between one to 20 milliseconds, which is still very short in comparison with the recovery times of cloud and bubble chambers.

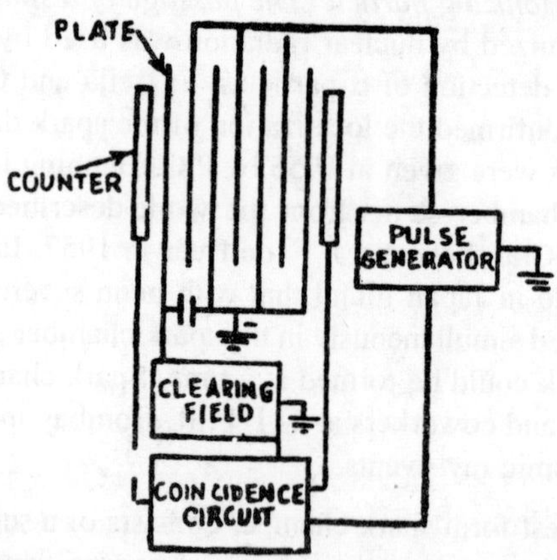

Fig. 4 : Schematic representation of spark chamber circuit.

The chamber can be operated in a magnetic field and the momentum and charge sign of the ionizing particle can then be determined from the curvature of the track. The number, thickness, spacing and material of the plates in a spark chamber are varied to suit the experimental requirements. The plates are commonly made of very thin aluminium to minimize energy absorption from the particle and are closely spaced to define the track more accurately. The sparks tend to be perpendicular to the plates in their passage between the adjacent plates. The photograph of spark chamber tracks. Sometimes it is desired to study the interactions of specific particles with matter. The plates are then thicker usually and are made of either carbon or a metal, such as iron, lead, tantalum or brass. The interactions take place almost within the solid plates as the density of nuclei in the gas between the plates is small.

One of the main advantages of the spark chamber over the other chambers is that the triggering and removal of ions by the clearing field are comparatively simple. With the help of this chamber it is possible to study an individual event produced by one particle among the many passing through the chamber at a rate that may be as high as a million per second. The spark chamber event of interest is selected by an arrangement of counters, called a *Logic circuit.*

Analysis of spark chamber photographs can be carried out by ordinary calculations or by computer techniques. The preliminary scanning stage may not be required because spark chamber events of interest have been selected by the logic system. The tracking, geometry and kinematic analysis arc performed in the manner as with bubble chamber pictures.

There are *two main drawbacks* to a spark chamber as compared with a bubble chamber. First, in the bubble chamber the liquid may be chosen so as to restrict the interactions of the incoming particle to those of a special type. Second, the co-ordinates of tracks and of the points of interaction in a bubble chamber can be defined with much greater accuracy than is possible in a spark chamber. As a general rule, bubble chambers are preferred where good space resolution is desirable. On the other hand, spark chambers permit the selection of events of interest with a high repetition rate, *i.e., they have better time resolution.*

Since 1963, various "filmless" spark chambers have been developed in which photography of the spark discharge is avoided. In such devices, the co-ordinates of the spark tracks are obtained directly in a form suitable for a computer to determine the geometry of the event. The grounded plate of every pair is replaced by an array of closely spaced parallel wires. For the high voltage plates a similar arrangement of wires could be used but solid plates are generally employed since they are more convenient and easier to fabricate. The wires arc made from a magnetostrictive material (50% Fe and 50% Co alloy) in *one type of wire spark chamber*. A passage of a spark from a particular wire produces a local magnetic field and a deformation occurs at the affected point. The deformation travels then along the wire to suitable detectors at each end

The ratio of the times between the firing of the spark and arrival of the deformation at the ends gives one co-ordinate, the other co-ordinate of the spark track is obtained by identifying the particular wire at which signals were received. The most promising wire spark chamber, referred to as a *spark-chamber hodoscope*, differs from the acoustic wire chamber in respect that the wires constituting the alternate grounded plates are alright angles to each other. At one end each wire is threaded through a ferrite core, such as is used in computer memory devices. A ferrite material has two stable states of magnetization and by a magnetic field (from an electric current) either one can be converted to the other. The magnetism of ferrite cores associated with the wires from which the spark passes is changed from one state to the other by the spark current or we can say that ferrite cores are nipped. By interrogating the cores in a manner similar to that employed in a computer the locations of the affected wires are determined and flipped cores are reset at the same time. Since successive planes of wires are at right angles, the information from the cores can give the co-ordinates of the spark track in three dimensions in digital form. It has a advantage of a very short recovery time. The amount of energy required to flip the core is very small and a feeble spark, which would not give off enough light to be photographed, is adequate for this purpose.

Sometimes it is desirable to make observations of interactions at very-very short intervals with high energy particles. *The scintillation counter hodoscope* can be used for this purpose. Although the accuracy of determining the co-ordinates of particle tracks decreases. It consists of an array of long thin rectangular strips (25 × 0.5 × 0.5 cm) of a plastic scintillator material. These strips are set side by side to produce a sensitive planer area. A photoelectric tube is associated with each strip and the passage of a charged particle through a given strip is indicated by the output signal from the corresponding photo-tube. This provides one co-ordinate of the particle track. For the determination of co-ordinates in the other two directions, it is necessary to build a hodoscope from several layers of scintillators arranged with the strips in successive layers at right angles. The output signals can be fed directly to an on-line computer for analysis. This type of hodoscope can record events at a much faster rate than wire spark chamber hodoscope can. It is because of the very short recovery time of a scintillation counter. The accuracy with which the track co-ordinates can be determined in a scintillation counter hodoscope is limited as it is dependent on the width of the scintillator strip.

DEKATRON

An alternate method of counting is provided by the multi-electrode cold cathode scaling tube known as dekatron. It is designed so a glowing spot moves successively around its circumferences with successive input pulses. It is a neon filled (13 cm. Hg pressure) tube containing an anode in the form of a flat disc of diameter ~1 cm. Surrounding the anode are thirty identical electrodes. They are symmetrically spaced and are about 0.05 cm. from the anode. These are grouped into three categories based on the operation principle of the device. Ten of the electrodes act as cathodes, A, to A, connected to a common point and K_0 the output cathode. Adjacent to each cathode is a pair of identical electrodes known as transfer electrodes or guides. The electrode nearest the cathode in a given direction is labelled guide 1 or 1G and the next electrode in the same direction is labelled guide 2 or 2G. The ten 1G electrodes are connected to a common terminal as are the ten 2G electrodes. The guide 1G and 2G electrodes are biased 20 V above the potential of the cathodes so that normally the discharge rests on one of the cathodes. Switch is normally closed but is opened for resetting to zero.

When the switch is open, K_0 is the most negative electrode and the gas discharge occurs between K_0 and the anode. With switch closed and a large negative input pulse applied to 1G, the slow discharge transfers from K_0 to $1G_0$. The duration of pulse should be greater than 60 μ sec. to ensure transfer. A delayed negative pulse is applied to the 2G electrode so that $2G_0$ remains negative after $1G_0$ returns to its normal bias potential of +20 V and the glow discharge moves smoothly to $2G_0$. The subsequent removal of the negative pulse from $2G_0$ moves the discharge to the next closest and most negative electrode K_1 and thus lights up the number 1 on the face plate. A second pair of negative pulses moves the discharge to $1G_1$, $2G_1$ and then to K_2 where by the number 2 is illuminated. Thus we see that the net result of a single drive pulse is to transfer the discharge from one cathode to the adjacent cathode. Two guide electrodes are required to ensure that the discharge always transfers to the next digit cathode and does not jump back. The current for the discharge is supplied by the anode voltage source. The input pulses are used only to stimulate the movement of the discharge from one cathode to the next. Every tenth input pulse transfers the discharge from K_9 to K_0. A load resistor R_{K0} produces a positive output or carry pulse. The carry pulses are amplified to produce large negative drive pulses for the next Dekatron stage.

Direct readout of the count can be accomplished within each tube by placing numerals around the periphery, adjacent to the glow discharge positions. Other techniques include stacking discs with number shaped perforations to makeup the common anode and by forming each cathode into the shape of a number.

IONIZATION CHAMBER

An ionization chamber, in its simplest form consist of vessel usually field with gas and into which it inserted an insulated electrodes. The gas becomes ionized when charged particles pass through it. If no voltage is applied to the container there will be no directing force acting the ions and no current will flow through the gas. If we applied a small potential difference to the electrodes, the –ve ions (or electrons) will be attracted to the anode and the +ve ions will go to the cathode. At the surface of the anode there will be a high concentration of negative ions. Recombination in this region will be a small because the rate of recombination is proportional to the product n^+n^- and here n^+ is very small.

All positive ions produced in this region move out promptly towards the cathode. Therefore, the recombination will be maximum at the center of the space. A small current will flow through a resistor connecting the electrodes externally As the voltage between the electrodes is increased, the ions will move faster, the number of recombination will be minimum and the pulse size will thus be increased. The pulse size increases with the voltage, until it reaches to a value corresponding to the no recombination of ions. This value of current is called the *saturation current* and the chamber is said to reach in *saturation.*

The ionisation chamber works in the region of constant pulse size. In this region the applied voltage is high enough to prevent recombination of ions and low enough to prevent multiplication.

In the initial stages, the gold-leaf electroscopes of great sensitivities or the quadrant electrometers were used. The next step in the development was a modification of the electroscope, designed by Lauritsen, where the gold leaf was replaced by a tiny metal coated quartz fibre. It could not give quick response to particles that follow one another in rapid succession. This difficulty was overcome to a great extent by a device known as linear or dc amplifier, which increases, without distorting, the feeble ionisation current from a single particle to a value which can easily register on an oscillograph or an electrical counter, dc amplifiers are more susceptible to disturbance and drift and more difficult to arrange with several successive stages of amplification than those designed for amplifying alternating currents.

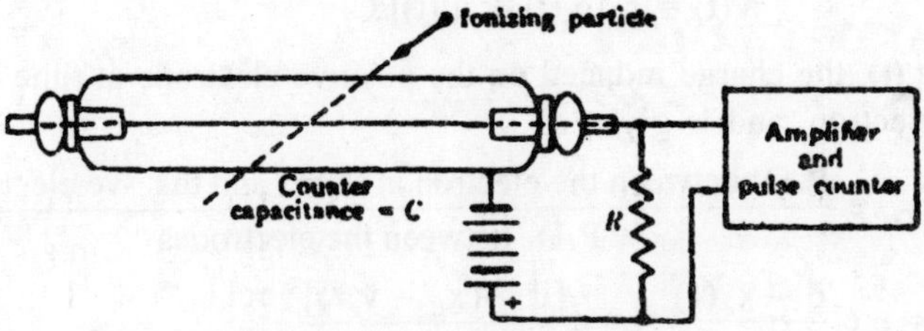

Fig. 5 : Ionization chamber.

Ionization chamber instruments may be grouped into two categories, depending upon the value of the time constant RC of the system relative to the frequency of arrival of the ionizing events. If RC is long compared to time between ionizing events a steady state is reached and a *direct current* may be measured. On the other hand, if RC is small each pulse produced by an ionizing event may be detected separately. These two types are also known as *integrating and non-integrating* respectively and the instrument as current and pulse ionization chambers.

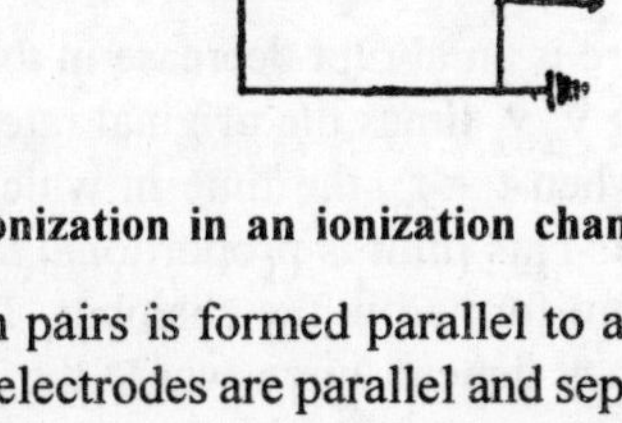

Fig. 6 : Ionization in an ionization chamber.

Let a track of n ion pairs is formed parallel to and at a distance x_0 from the axial wire The electrodes are parallel and separated at a distance d. A voltage V_0 is applied to the electrodes, so that electrons and positive ions drift in opposite directions in a uniform electric field V_0/d. Let us first assume that the time constant RC (where C = capacitance of the chamber + input capacitance of the amplifier) is much greater than the collection time of ions. So that the current through R may begin as soon as the electrons and +ve ions move apart in the chamber.

The change of potential at the positive electrode at the time t, due to a track of n ion pairs formed at t = 0 is given as

$$V(t) = n\,[q_e(t) + q_p(t)]/C, \quad ...(1)$$

where $q_e(t)$, the charge induced on the positive electrode at time t by a single electron, and is given as

$$q_e(t) = -e.\frac{\text{P.D. between the electron at time t and the -ve electrode}}{\text{P.D. between the electrodes}}$$

$$= -e.\frac{d - x_e(t)}{d} = -e\frac{[d - (x_0 - v_e t)]}{d} \quad ...(2)$$

and $q_p(t)$, the charge induced on the positive electrode at time t by a single positive ion, is given as

$$q_p(t) = +e.\frac{\text{P.D. between the +ve ion at time t and the -ve electode}}{\text{P.D. between the electrodes}}$$

$$= e.\frac{d - x_p(t)}{d} = e\frac{[d - (x_0 - v_p t)]}{d} \quad ...(3)$$

The relations (2) and (3) are only true for parallel plate geometry in which potential gradient is uniform.

From eq. (2), (3) and (1) we get.

$$V(t) = -ne(v_e t) + v_p t)/Cd. \quad ...(4)$$

As the electron drift velocities v_e are about a thousand times greater than the heavy ion velocities v_p, hence, for the time between 0 and t_e is the electron collection time ($x_0 = v_e t_e$), the above relation transform to

$$V(t) = -nev_e t/Cd. \quad ....(5)$$

For $t > t_e$, there is an abrupt decrease in the rate of rise of potential, the new rate being v_p/v_e times the original rate. The change of potential becomes—ne/C when $t = t_p$, the time in which positive ions reach the negative electrode. This limit is proportional to n and is independent of where the ions were formed in the chamber. The shape of the collector potential curve does depend upon where the ions are formed.

With sufficient amplification, a large pulse will appear at the amplifier output terminal. To have the height of the output pulse proportional to the amount of ionization produced by the particle in the chamber, two kinds of proportional or linear amplifiers can be used. One, known as slow *amplifiers*, in which the shortest time constant is chosen long compared with the drift time of the +ve ions, have several disadvantages.

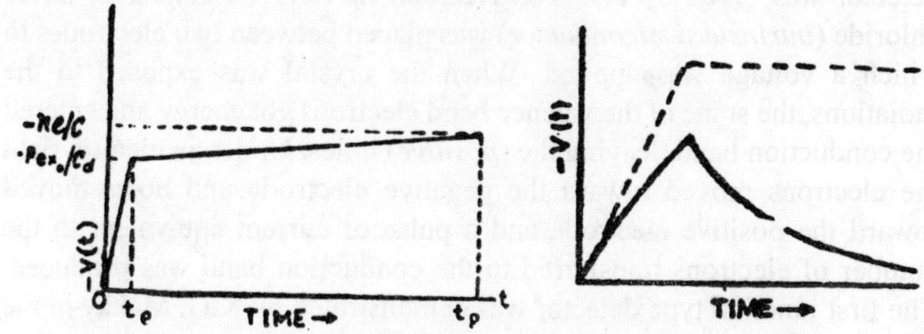

Fig. 7 : (a) Shape of potential pulse. (b) Pulse cliping.

The others, known as fast amplifiers, in which time constant RC is made short enough so that the induced potentials by the +ve ions are not of interest and clip the pulses Thus the slow linear rise, after the collection of electrons, is removed and the height of the recorded pulses depends on $V(t_e)$ and not on $V(t_p)$. If the complete range of an *ionizing particle is confined to the region between the negative electrode and grid*, the grid screens the +ve electrode from the electrons and +ve ions. Electrons, while drifting through the grid, induce a change of potential at the +ve electrode. As positive ions are screened from the +ve electrode by the grid, hence have no effect. Thus, the pulse height increases due to the induction effect of the moving electrons. When all the electrons are collected it attains a maximum value given by

$$V_{max} = -\frac{ne}{C} \quad ...(6)$$

It is also independent of the position and orientation of the ionization track.

The height of the amplified pulses from the ionization chamber

$$V = -\frac{AEe}{WC} \quad ...(7)$$

where A is the voltage gain of the amplifier, E the energy of the ionizing particle, e the electronic charge and W the mean energy required to produce an ion pair in the gas filled in the chamber.

SOLID-STATE DETECTORS

Crystal counters are essentially the ionization chambers using solid dielectrics between parallel plate electrodes. The first solid state nuclear detector was given by P.J. Van Heerden in 1945. A crystal of silver chloride (*intrinsic semiconductor*) was placed between two electrodes to which a voltage was applied. When the crystal was exposed to the radiations, the some of the valence band electrons got energy and entered the conduction band, leaving the (*positive*) holes. Under an electric field the electrons moved toward the negative electrode and holes moved toward the positive electrode and a pulse of current equivalent to the number of electrons transferred to the conduction band was produced. The first junction type detector was demonstrated by K.G. Mckay in the United States in 1949. A collimated beam of α-particles was allowed to fall on a n-p-germanium junction. He showed that nearly whole of energy of *α-particles absorbed in it was utilized in producing electron hole pairs and the average energy required to produce one electron hole pair was about 3eV which was very small in comparison to 34eV, the energy required to form an ion pair in a gas.*

J.D. Mayer and B. Gossick showed that *the size of the output pulse was proportional to the alpha-particle energy. In this respect, a solid state detector behaves like an ideal pulse type ionization chamber*. Most semi-conductor junction type detectors fall into three categories, according to the formation of junction.

1. Diffused Junction Detector

In this detector, a thin layer of the n-type is formed on the surface of a p-type silicon (or *germanium*) crystal, by allowing a donor impurity (*phosphorous*) to diffuse into one face of the crystal at a fairly high temperature. There is a tendency for electrons to diffuse from the n-region (*where they are in excess*) into p-region (where the electrons are in deficit). An equilibrium is reached when the fermi-level is constant throughout the material. Due to this an intermediate volume around the interface will be cleared from carriers of both signs and this region is called the *depletion layer*. This layer is of low conductivity and high field, hence offers ideal conditions for the detection of ionizing radiation.

When the negatively charged electrons diffuse away from the n-type region, they leave the material positively charged. Similarly, the filling of the hole in the p-region generates a corresponding negative charge. The small potential difference of about a volt, established in this manner, is enhanced by an additional potential of several hundred volts, referred to as a reverse-bias voltage. In the presence of this electric field the positive holes in the p-type side are pulled towards the –ve electrode, the electron towards the p-n junction and thus, a depletion layer is created in the p-type material. The result is an extremely low leakage current. The thickness of the depletion layer increases with the reverse-bias voltage and the limit is set by electrical break-down in the semi-conductor.

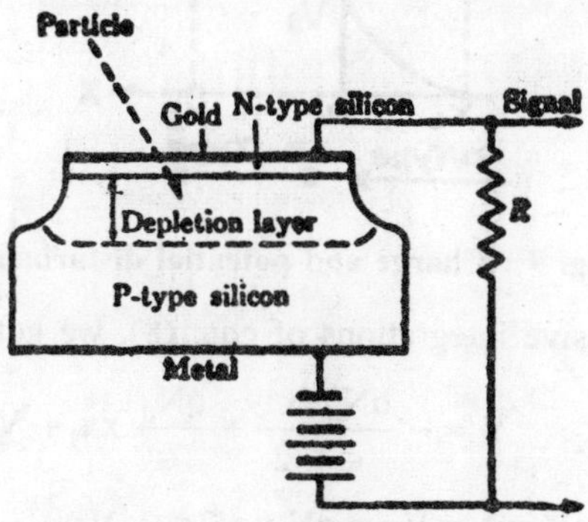

Fig. 8 : A p-n junction detector.

Thickness of Depleted Region

The thickness of the depleted region and the capacitance can be calculated by Poisson's equation. Assuming the variation of V in the direction ⊥ to the surface separating the n- and p-regions, we have

$$\frac{d^2V}{dx^2} + \frac{\rho}{\epsilon} = 0 \qquad ...(8)$$

ρ the volume charge density given by

$$\rho = q(p + N_{\&} - n - N_a) \qquad ...(9)$$

where p, n, $N_{\&}$ and N_a denote, respectively, the concentrations of holes, electrons, donors and acceptors. In the completely depleted region n and p are negligible, hence we have

$$\rho = q\,N_{\&} \quad 0 < x < x_1$$

$$\rho = -q\,N_a \quad -x_2 < x < 0$$

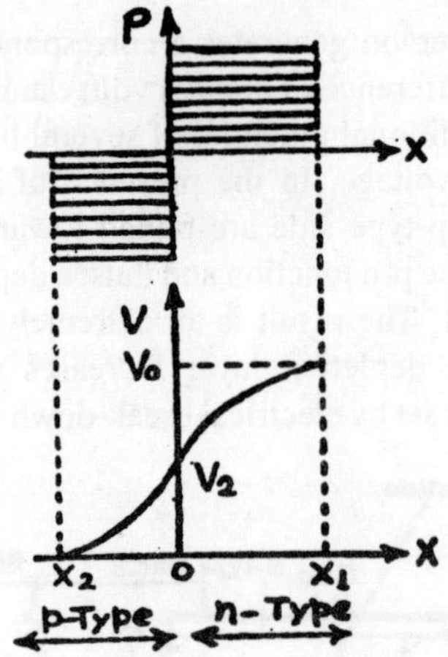

Fig. 9 : Charge and potential disturbance.

After successive integrations of eqn. (8), we get

$$V = -\frac{qN_d}{\epsilon}\frac{x^2}{2} + \frac{qN_d}{\epsilon}xx_1 + V_2$$

$\therefore \quad V(x_1) = V_0 = qN_dx_1^2/2\epsilon + V_2$

or $\quad V_0 - V_2 = qN_dx_1^2/2\epsilon$ ...(10)

Similarly $V(-x_2) = 0$ and $V_2 = qN_a \dfrac{x_2^2}{2\epsilon}$ ...(11)

Hence the potential barrier height becomes

$$V_0 = q(N_dx_1^2 + N_ax_2^2)/2\epsilon \quad \text{...(12)}$$

As charge is conserved, hence $N_dx_1 = N_ax_2$ and

$$V_e = \frac{q}{2\epsilon}\left(1 + \frac{N_a}{N_d}\right)N_ax_2^2 = \frac{q}{2\epsilon}\left(1 + \frac{N_d}{N_a}\right)N_dx_1^2$$

The thickness of the depletion layer

$$X_0 = x_1 + x_2 = \left[\frac{2\epsilon V_0}{q}\frac{N_a + N_d}{N_aN_d}\right]^{1/2} \quad \text{...(13)}$$

If p-type region is more heavily doped, than the other region, we have $N_a > N_d$ and $x_2 < x_1$. Hence, we have

$$X_0 \simeq x_1 = \left[\frac{2 \in V_0}{q N_d}\right]^{1/2} = [2 \in \rho_n \mu_n \ V_0]^{1/2} \qquad \text{...(14)}$$

where ρ_n is the resistivity of n-type region and μ_n the electron mobility.

The same treatment will give the value of X_0 in the presence of reverse bias, as

$$X = \left[\frac{2 \in (V_0 + V)}{q N_d}\right]^{1/2} = [2 \in \rho_n \mu_n \ (V_0 + V)]^{1/2}$$

If the applied voltage $V > V_0$, we get

$$X \simeq [2 \in \rho_n \mu_n \ V]^{1/2}, \qquad \text{...(15)}$$

where a is a constant of a medium.

The depletion layer having area A has a capacitance C given as

$$C = \frac{\in A}{X} = A\left[\frac{q \in N_a N_d}{2(N_a + N_d)(V_a + V)}\right]^{1/2}$$

For $N_a > N_d$ and $V > V_0$, we have

$$C \simeq A\,(\in/2\ \rho_n \mu_n\ V)^{1/2}. \qquad \text{...(16)}$$

If ρ is in ohm cm, V in volts, X in microns (μ), and C in pf/cm² then we have following values:

G_e	n-type	$X = (\rho V)^{1/2}$	$C = 1.37 \times 10^4\ (\rho V)^{-1/2}$
	p-type		
		$= 0.65\ (\rho V)^{1/2}$	$= 2.12 \times 10^4\ (\rho V)^{-1/2}$
Si	n-type	$X = 0.5\ (\rho V)^{1/2}$	$C = 2.2 \times 10^4\ (\rho V)^{-1/2}$
	p-type	$= 0.3\ (\rho V)^{1/2}$	$= 3.7 \times 10^4\ (\rho V)^{-1/2}$

These detectors should always be kept in complete darkness, otherwise the photosensitivity will bring out a *photoelectric current.*

2. Surface Barrier Detector

It has been very useful for charged particles detection. It is made by exposing a chemically etched surface of an n-type silicon crystal to air. The etched surface is thus oxidised and this oxidation laye-acts like a very thin p-type layer.

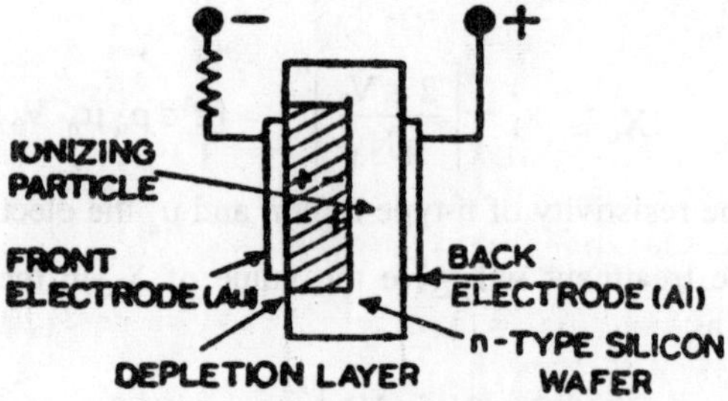

Fig. 10 : Surface barrier detector.

In analogy to an N-P junction, a depletion layer is formed, but it is clear that this layer only exists with in the semiconductor. The thickness of the depletion layer X_0 (or X with the reverse bias V) and the capacitance can be written as given in the case of N-P junction. Using 4000Ω silicon with the reverse bias of 1000 volts, this thickness reaches to 3 mm. In contrast to N-P junction, the temperature attained during the various stages of the construction of the surface barrier detectors does not exceed the boiling point of water, and the life time remains as that of the original material.

Detection Characteristics of Surface Barrier Detectors

(a) *Resolution:* The high energy resolution means a narrow distribution of pulse size around the mean, *i.e.,* FWHM (full width at half maxima). With monoergic radiations,, Siffert, *et al.*, found that the value of FWHM was smaller for protons and deuterons than for α-particles. The broadening of peak for α-particles was due to nuclear collisions. The FWHM is found to be ~10 keV for α-particles of several MeV. Mayer and others suggested FWHM of 2.1keV at 135°K. The cooling increases the resolution. It is thus usually necessary to cool the detector to –30°C.

(b) *Linearity:* Many experiments verify the linearity of the response to various particles with different energies. In addition to the defect related to a thin space charge region, other deviations arise for heavy particles and fission fragments.

(c) *High Conversion Efficiency:* The energy required to produce ion pair in silicon is 3.5eV. Thus, the number of pairs for a particular incident energy is about 10 times the number of pairs in air.

(d) *Differential Sensitivity:* The surface barrier detectors are relatively insensitive to neutrons or photons. They can detect charge particles in the presence of other radiations.

(e) *Small Size:* These detectors are of very small size and are important. In some cases the small sizes are a drawback.

(f) *High Speed of Response:* The carrier mobilities are high (~1500cm^{2}volt/sec) for electrons and 500 for holes. The ion pairs move only a short distance before the collection and the pulse widths are in the order of nanoseconds.

3. Lithium Ion Drifted Junction Detector

The above limitation was reduced by E.M. Pell in the United States in 1960 after introducing the lithium-ion drifted junction detector. The technique is to allow the lithium, which acts as an electron donor, to diffuse into the p-type silicon or germanium at an elevated temperature (120 to 150°C) and under reverse bias. Under these conditions the ions drift for a considerable distance into the crystal where the lithium donor exactly compensates the existing acceptor impurities in the p-region and an effective intrinsic semi-conductor layer is produced between the n- and p-regions.

Principle and Characteristics

The first step of Pells method is to form an N-P junction, by diffusing Lithium from a layer deposited on the surface of a p-type material. If a reverse bias V is applied to this diode, +vely charged Li-ions will move from n-region to p-region. The electric field E is assumed to be large enough to make drifting much more important than the diffusion. To obtain a sufficiently large mobility μ, the temperature of the material may be reached. In the case of a long drift time, the thickness

$$X \simeq (2\mu\ V^t)^{1/2} \qquad ...(17)$$

In the presence of a reverse bias, the electrons move towards N-region and holes towards P-region. A space charge is thus produced which varies linearly with distance, if recombination is neglected (Fig. 11b). This space charge causes a distortion of the electric field. The field becomes minimum at the centre of the intrinsic region (Fig. 11c), when the mobilities are equal. The distribution of Li-ions must be such that the space charge is compensated in the intrinsic region. It varies with distance as shown in (Fig. 11c). When the diode is cooled after the drift the thermally generated carriers become very small but the lithium distribution remains unchanged.

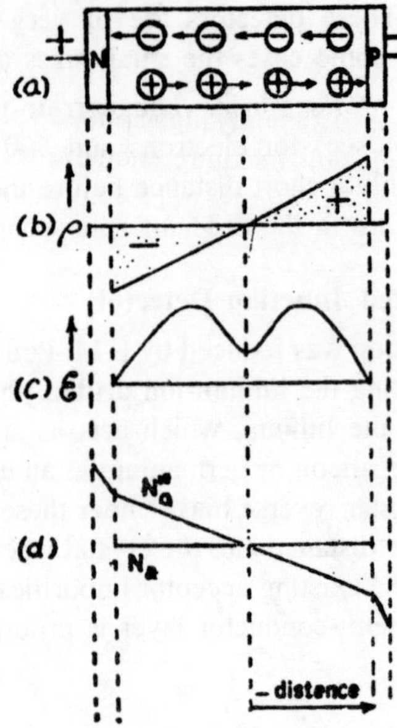

Fig. 11 : NIP-diode.

Lithium-drifted germanium detectors are more suitable than silicon detectors for the detection of γ-rays. It may be recalled that the photo-electric absorption cross section for γ-rays is proportional to Z^5 and therefore germanium is more efficient than silicon for γ-ray

Advantage

1. The elimination of windows and the requirement of low applied voltage to operate the counter are advantages over the gas counters.
2. These are more suitable than the grid ionization chambers coincidence experiments.
3. Very good energy resolution is possible.
4. Counting rates high as 5×10^4 counts/sec. are possible.
5. If the particle range is less than the junction thickness, the pulse height is proportional to the K.E. of the incident protons, deuterons, α-particles or any other ionizing particles.

6. These detectors have low sensitivity to gamma background radiation.

REGIONS OF MULTIPLICATIVE OPERATION

Ion current or pulse height in an ionization chamber increases above the saturation value at sufficiently high applied voltages because electrons moving in these high fields acquire enough energy to cause secondary ionization. The size of the current pulse depends on the number of initial ion pairs produced by the ionizing particles and on the applied voltage. For a given ion pairs, the pulse size is shown in Fig. 12 on a logarithmic scale as a function of the applied voltage.

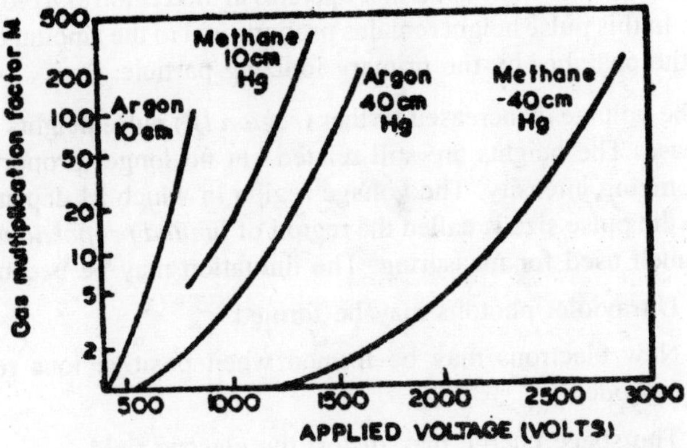

Fig. 12 : Variation of pulse size (log) with applied voltage.

By these results shows that the curve can be divided into six more distinct regions.

In region I, the voltage applied is so low that recombination of ion pairs takes place. As the voltage is increased there is a competition between the loss of ion pairs by recombination and the removal of ions by collection on the electrodes. The region is known as ion recombination region,.

In region II, the voltage is sufficiently high so that only a negligible amount of recombination takes place and the ion pairs move to the electrodes so rapidly that virtually every ion pair reaches the electrodes.

Thus, the pulse size remains independent of the potential as long as the voltage is increased upto the point at which secondary electrons arc produced by collision in the gas. This region is known as *ionization cham-ber region or saturation region.*

In region III, voltage is sufficiently high so that electron approaching the central wire attains suffi-cient energy to produce new ions, due to collision with gas molecules. The original electron and the new one, thus, produced, are again accelerated by the field and more ions are still produced by collision. This process is cumulative and an avalanche of electrons is thus produced. The phenomenon fit frequently called the *Townsend avalanche*, in honor of J.S. Townsend who did pioneer work in this subject. The number of the ion pairs formed by collision as each electron travels toward the central wire is known as *multiplication factor M.* The voltage region in which M remains constant is called the *proportional region.* The apparatus operates in this region as a proportional counter. In this pulse height remains proportional to the amount of energy lost in the chamber by the primary ionizing particle.

If the voltage is increased further (*region IV*) pulse heights continue to increase. The heights are still related but no longer proportional to initial ionizing intensity. The voltage region in which M depends some what on the pulse size is called the region of *limited proportionality* and is not much used for measuring. The limitation may be because:

(a) Ultraviolet photons may be formed.

(b) New electrons may be formed when positive ions reach the cathode,

(c) The space charge may distort the electric field.

For still higher voltages the *region V* is reached, in which pulse height is very large and also substantially independent of the initial number of ions. This is called *Geiger region.* The device operated in this region is called a Geiger-Müller counter. In the Geiger region the avalanche associated with the drift of a single electron to one point on the anode initiates avalanches over the entire anode length.

At the end of the Geiger region the counter goes into continuous discharge region,

PROPORTIONAL COUNTER

In the parallel plate ionization chamber if the electrode voltage is increased by beyond the saturation interval up the primary ions produced secondary by collisions with the gas: consequently, the primary ionization is multiplied by a factor that depends on the Geometry of the apparatus and on the applied voltage v the chamber under these condition is ion own as proportional counter.

For a cylindrical geometry, the electric field required to produce secondary ionization is given by

$$E = \frac{V}{r \log_e b/a}$$

Where,

a = radius of central wire

b = inner radius of cathode cylinder

V = electrode or central wire voltage

r = distance of electric field E from the central wire.

The electrons ejected from gas molecules by ionizing particle drift towards the positive electrode. If they move in a region of high electric field E, they will produce additional ionization by collisions with molecules of the gas. In drifting a short distance ax, n electrons will increase the number of elections to n + dn, where

$$dn = \alpha \, n \, dx \qquad ...(18)$$

Here a is the first *Townsend coefficient*. If n_0 is the number of electrons set free by ionizing particle and x is the distance through which these electrons drift to reach the positive electrode, the number of electrons reaching this electrode is given by

$$n = n_0 \exp.\left[\int_0^x \alpha \, dx\right] \qquad ...(19)$$

The ratio n/n_0 is called the *gas amplification or gas multiplication factor M.* The radius of the cylinder and a the radius of an axial wire, the electric field strength at a radius x is

$$E_x = \frac{V_0}{[x \log_e b/a]} \qquad ...(20)$$

Let us assume that every electron produced in the primary ionization gives rise to n-secondary electrons by collisions. The production of secondary electrons during the collision is associated with the photons which in turn produce photoelectrons. If p is the probability of production of photoelectrons, there will be np photoelectrons in all. If each photoelectron produces n-electrons by further collisions, we shall have second-generation avalanche of n^2p electrons, which in turn will give rise to more electrons.

$$\therefore \text{ Total number of electrons } M = n + n^2p + n^3p^2 + ... \qquad ...(21)$$

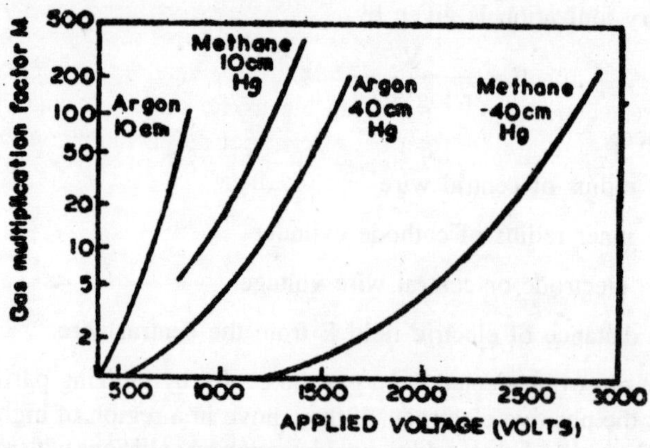

Fig. 13 : Variation of M with applied voltages.

For practical proportional counter $np < 1$, hence $M = n/(1 - np)$. It represents that in a proportional counter the total number of secondary electrons is proportional to the number of initial electrons. The variation of M with voltage in argon and methane at two different pressures is shown in the given Fig. 13.

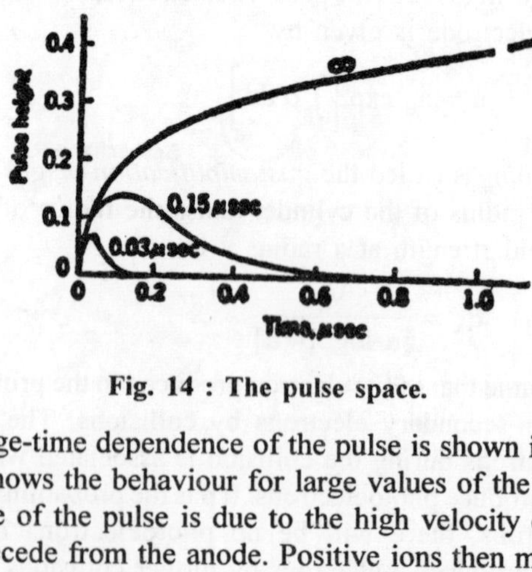

Fig. 14 : The pulse space.

The voltage-time dependence of the pulse is shown in Fig. 14. The upper curve shows the behaviour for large values of the time constant. The initial rise of the pulse is due to the high velocity of the positive ions as they recede from the anode. Positive ions then move in a weak field and the pulse approaches its final height extremely slowly. For the case of a finite time constant the pulse has the shape shown by lower curves. In practice, suitable clipping time constants (RC ~ 0.1μ sec.) are used in the output circuit so that better time resolutions are available.

If the primary ionizing particles arrive in rapid succession, mechanical counter can not record all the pulses. The output from the amplifier is therefore first passed through a scaling circuit before being fed into a mechanical counter. A sealer is an electronic circuit arrangement that feeds a predetermined fraction of the actual particle counts to the mechanical counter.

GEIGER-MÜLLER COUNTER

Geiger and Müller in Germany developed the counter in 1928. This counter is thus named as Geiger-Müller counter or G.M. counter. It works in Geiger region. The main difference between proportional and Geiger counter is that in the *former the avalanche is formed only at one point whereas in the latter the avalanche spreads along the whole length of the central wire. The amplification does not therefore depend on the initial ionization produced by the ionizing particle.*

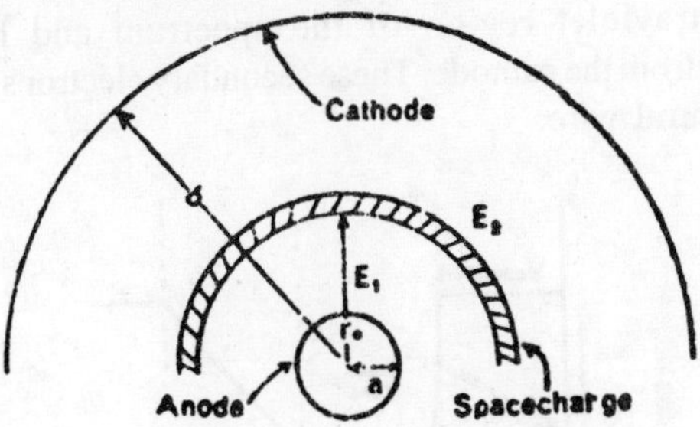

Fig. 15 : Positive ion sheath.

Pulse Formation and Decay

Let us Consider the situation when the space charge has moved outside to a radius r_0. Let there be q charges per unit length on the wire and Q charges per unit length in the space charge. The field strengths inside and outside the sheath are

$$E_1 = q/2\pi\epsilon r \text{ and } E_2 = (q + Q)/2\pi\epsilon r.$$

Voltage applied to the counter

$$V = \int_a^{r_0} \frac{q}{2\pi \in r} dr + \int_{r_0}^{b} \frac{q + Q}{2\pi \in r} dr$$

Integrating we have

$$= \frac{q}{2\pi \in} \log_e \frac{b}{a} + \frac{Q}{2\pi \in} \log_e \frac{b}{r_0}$$

$$\therefore \text{ Charge} \quad q = \frac{2\pi \in}{\log_e b/a}\left[V - \frac{Q}{2\pi \in} \log_e \frac{b}{r_0}\right] \quad ...(22)$$

The negative term in the bracket shows that q and hence E_1 is reduced during a pulse. As the positive ion sheath moves toward the cathode, r_0 increases, negative term decreases and the field strength attains its maximum initial value. The positive ion sheath will reach the cathode in about 10^{-4}sec. As the positive ions reach the wall and are about 10^{-7}cm from it they cause electron to be liberated from the metal of the cathode by *field emission*. These electrons then neutralize the positive ions, leaving the excited molecules. From where they promptly go to the ground state radiating characteristic spectral lines. Some of these lines are in the ultraviolet region of the spectrum and hence liberate photoelectrons from the cathode. These secondary electrons are accelerated toward the central wire.

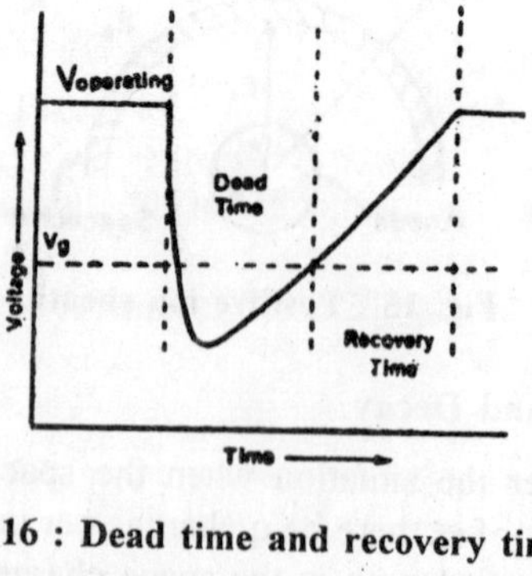

Fig. 16 : Dead time and recovery time.

When the electrons reach the wire, they start a new avalanche which produces a new sheath and the discharge will continue indefinitely. This continuous discharge is not really continuous in time but consists of a series of pulses at close and some what overlapping intervals. The growth of the potential pulse at the anode is determined by the motion of the

+ve ion sheath as it is repelled away from the high field region. When the ion sheath has travelled a distance equal to about half the radius of cathode cylinder, the field close to the anode wire is below the Geiger threshold. Hence the second pulse can not be recorded. The time, during which the counter is incapable of responding to a second ionizing event, is known as *dead time* (~200μ sec.). As the space charge moves further towards the cathode, the counter sensitivity gradually returns to its original value and the second pule can now be recorded.

This time, during which pulses of reduced size are produced, is known as *recovery time.* The dead time plus recovery time may corresponds to the time which +ve ion sheath takes to reach the cathode, and is known as the *resolving time* of the counter. The high counting rate is generally reduced by reducing the high potential supply to the counter for a definite time Interval called *paralysis time*. It is achieved by an external electrical circuits generally a relaxation oscillator triggered by the step rise of Geiger pulse;

Quenching of the Discharge

The process used to prevent a continuous series of pulses from taking place in the G.M. counter are called quenching. There are three types of quenching (*suppressing*) involved in the operation of a counter:

(a) quenching of photons in the initial avalanche,

(b) electrostatic quenching of the avalanche by the +ve ion space charge, and

(c) quenching of the secondary emission when the +ve ions reach the cathode. The quenching of the discharge involves both the electrostatic quenching and the quenching of secondary emission. There are two procedures for quenching the discharge in order to improve the resolving power of G M. counters.

In the *nonself-quenched* (externally quenched) type of G. M. counter, the quenching of the discharge is achieved by means of an external resistance or by use of an auxiliary electronic circuit. In the first method counter can be quenched by using the load resistor ($10^{10}\Omega$) in series with the tube. High voltage is applied across the tube which reduces as the current pulse flows through the resistance, the gas amplification factor is thus decreased preventing a second discharge. With a chamber of 10pf capacitance this method results in a counter dead time of the order of 0.1 second. In the second method the load resistance is about $10^7\Omega$ and at the same time the voltage across the chamber is reduced, by the application of a negative voltage pulse to the anode from a monostable multivibrator, every time a count takes place. The dead time of the counting circuit can be adjusted electronically to be a chosen value.

In the *self-quenched* (internally quenched) type of G.M. counter, quenching is achieved by the inclusion of a suitable gas in the counter, such as:

(a) organic molecules, and

(b) the halogen gases. In the first case the filling is a mixture of argon and 10% of a polyatomic organic gas or vapour such as methane, ethane or ethyl alcohol as a quenching gas. As the quenching gag ionizes more readily than argon due to its lower (11.3 eV instead of 15.7 eV) ionization potential, hence the ion cloud reaching the cathode consists almost entirely of alcohol ions. These ions are neutralized by pulling electrons from the metal surface of cathode and again form largely excited molecules. *These excited molecules usually prefer decomposition rather than photon emission and hence no photons are available to start a second avalanche.* The photons if emitted by excited molecules are not likely to travel more than a millimeter through the gas before being absorbed by the photoelectric process. The process of charge exchange during collisions prevents any argon ion reaching the cathode, thereby preventing the release of any electrons from the cathode which would trigger off a second discharge.

The each transfer of ionization from argon to alcohol is accompanied by the emission of photons of energy 15.7 – 11.3 = 4.4eV. This is too low to eject an electron from a cathode.

A satisfactory quenching gas must nave three main properties:

1. When in an excited state it must prefer to dissociate rather than to de-excite by the emission of photons.
2. Its ionization potential should be lower than that of the main counting gas in the tube.
3. It must have broad and intense ultraviolet absorption bands.

The halogen quenching action seems to be similar to that of the organic agents in that some of the molecules dissociate at each discharge. However, the dissociated halogen molecules combine at the end of the discharge, so the tube has unlimited life. Halogen molecules are chemically so active that the materials of which the tube and its electrodes are made must be chosen carefully. The halogen quenched counters usually operate at 300 to 700 volts. They can be used over a temperature range of from –50°C to +60°C, whereas the organic quenched counters often cease to work at 0°C because the organic gas condenses. They are not damaged by a continuous discharge whereas poly-atomic organic gases filled counters damaged. Due to their lower voltages, long life and robustness, the halogen quenched counters are recommended for use in colleges.

In the self quenching type counters the discharge is quenched due to the internal mechanism and hence no electronic quenching circuit or high resistance is necessary. The disadvantage of high resistance is that it lengthens the recovery time of the counter. The equivalent electronic circuit permits a short time constant to be used but adds the inconvenience of an extra stage in the circuits. The advantage gained in the elimination of the quenching resistor or electronic circuit is secured at the price of stability.

G.M. Tube Construction

G. M. *tubes* have been made in a great variety of sizes and shapes, from 1 cm. to 100 cm. in length and from 0.3 cm. to 10 cm. in diameter. The walls can be of metal (copper), a metallic cylinder may be supported inside a glass tube, or the interior surface of the glass tube may be coated with a thin layer of an electrical conductor) *e.g.*, silver). The central wire is usually of tungsten with a thickness of 2×10^{-3} to 5×10^{-3}cm.

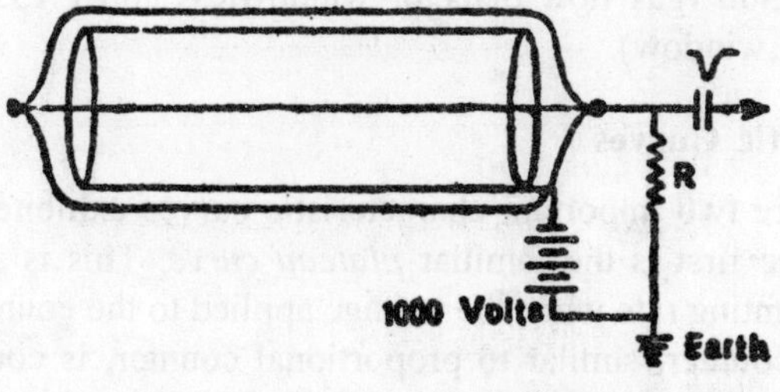

Fig. 17 : Basic circuit of G.M. counter.

Various observers have tried a large number of different compounds in counters and found that any polyatomic gas molecule would produce the quenching action. The molecule need not be organic. There does not appear to be any great difference in merit between various possible polyatomic molecules, each having certain advantages and disadvantages as compared to other. For example, amyl acetate shows a some what longer useful life but a slower resolving time than does methane. To extend the life time of the tube, bromine vapour can be introduced instead of the alcohol. To detect radiation of very small penetrating power counters with very thin end windows are used. Thin end windows of mica, cellophane or glass, only 1 or 2 mg/cm^2. thick will allow a- and P-particles to pass into the counter. For detecting P-radiation thin walled glass counters (the cathode is often a thin layer of colloidal carbon deposited on the inner wall of the glass envelope), which allow the entry of β-rays from all directions, can be used. Detection of γ-rays is mainly through the ejection of photoelectron from the cathode cylinder. The probability of this process varies as Z^4. hence, counters using high Z (bismuth or lead) cathode are used for the detection of γ-rays. For detecting cosmic rays more robust counters sometimes two or three feet long are used. For our purposes:

(i) *G.M. tube should be capable of responding to the radiation to be detected,*

(ii) *It should operate at fairly low voltages,*

(iii) *It should have long life, and*

(iv) *Its cost should be minimum.*

In India G.M. tube of different shape and size are designed in Hyderabad. These are marked as I 1000, I 1002 (Helogen filled); I 1010 (Thin walled); I 1030, I 1031, I 1032 (end windows); I 1200 (liquid sample); I 1300 (gas flow detector windowless and I 1350 (gas flow detector end window).

Characteristic Curves

There are two important characteristic curves exhibited by Geiger counters. The first is the familiar *plateau curve.* This is a plot of the recorded counting rate when the voltage applied to the counter is varied. The G. M. counter, similar to proportional counter, is connected with a device capable of indicating only relatively large pulses but no small ones. Until the voltage reaches the value indicated as the *starting potential* the pulses are too small to be detected. But with rising potential, the gas amplification increases and number of pulses rise rapidly to a flat portion of the curve called *the plateau.* This is the Geiger tube region for which the *count rate is nearly independent of the potential difference across the tube.* Beyond the plateau the applied electric field is so high that a continuous discharge takes place in the tube and the count rate increases very rapidly. It should be emphasized that the shape of the knee of the plateau curve is partly a property of the associated recording circuit.

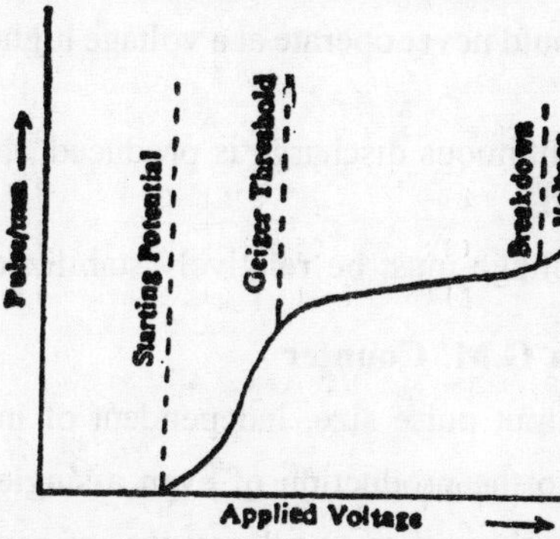

Fig. 18 : Plateau of G.M. tube.

$$\text{Slope of plateau } \frac{n_2 - n_1}{n_{av}} \times \frac{100}{V_2 - V_1} \times 100\% \text{ per 100 volts}.$$

The plateau has a width (or length) of a few 100 volts. A good tubes often have a plateau length of 100 to 200 volts or more with a slope of about 5% counts per 100 volts applied. *The slope for halogen filled tubes is however greater than that for organic gase filled tubes.* When the plateau becomes notably shorter and steeper, it is a sign that a counter tube is nearing the end of its useful life. The other type of curve is that of the starting potential as a function of the pressure of the gas in the counter. It can be shown from these curves that the *starting potential increases, practically linearly with the pressure of the filled gas.*

The *efficiency of the counter* is defined as the ratio of the observed counts/sec to the number of ionizing particles entering the counter per sec. Counting efficiency is defined as the ability of its counting if at least one ion pair is produced in it.

$$\therefore \qquad \text{Counting efficiency} = (1 - e^{-sl p}).$$

where s is the specific ionization at one atmosphere, p the pressure in atmospheres and *l* the path length of the ionization particle in the counter.

Precautions in Using G.M. Tube

1. The tube must be disconnected when measurements arc over.
2. Introduction of light must be prevented in order to avoid photo electric effects.
3. The background counting rate must be subtracted.
4. The tube should never operate at a voltage higher than its normal voltage.
5. In case a continuous discharge is produced, the voltage should be decreased.
6. The high voltage must be relatively stabilized.

Main Features of a G.M. Counter

(i) Constant output pulse size, independent of initial ionization.

(ii) Sensitivity to the production of even a single ion pair,

(iii) A long insensitive time to allow entry of each particle.

(iv) Infinite life (specially with halogen quenched),

(v) Can detect α, β, γ and cosmic rays.

Pulse Counting Circuits

The pulses obtained from the G.M. tube are too small to count direct. It requires amplification, which is done frequently in two steps. The counter is immediately followed by a *preamplifier*, which has little gain but produces an output signal across a relatively low impedance. This signal is then fed into the main amplifier, which has a gain strictly independent of the size of the input pulse. The output signal is then fed into the scaling circuit.

Resolving Time and Actual Counts

Let us assume that a counting system with a resolving time τ responds at a rate n counts per unit time when exposed to N initiating events per unit time. In unit time the total insensitive time will be $n\tau$ and the number of counts missed will be $Nn\tau$.

No of counts missed = error in counting or $Nn\tau = N - n$

$\therefore$ Actual count rate $N = n/(1 - n\tau)$

Thus, the actual count rate can be calculated by knowing τ. The most usual method for it involves the measurement of four counts using two radioactive sources. The background count (counts without any source) rate B is first found. With one source, count rate is determined. Let the observed count rate be $n_1 + B$ and expected $N_1 + B$. With the first, source the second source is introduced and adjusted until the count rate is approximately doubled. We now have $N_1 + N_2 + B$ expected and $n_{12} + B$ observed. The first source is removed and the second is left in position. If the counts rate is $N_2 + B$ expected and $n_2 + B$ observed. Solving equations

$$N_1 = \frac{n_1}{(1+n_1\tau)}, N_1 + N_2 = \frac{n_{12}}{(1-n_{12}\tau)} \text{ and } N_2 = \frac{n_2}{(1-n_2\tau)},$$

We get $$\tau = \frac{n_1 + n_2 - n_{12}}{2n_1n_2} \qquad ...(23)$$

Standard Deviation

Because of the random nature of radioactive decay it is found that if a long lived source is counted for a number of equal intervals of time the result shows considerable variations. This deviation is the square root of the average of the squares of the deviations from the mean. If N is the total number of counts in time t, then the

$$\text{Standard deviation } \sigma = \sqrt{\frac{(N)}{t}} \qquad ...(24)$$

$$\text{The coefficient of Variation } V = \frac{\text{Standard deviation } \sigma}{\text{Rate of counting R}}$$

$$= \frac{\sqrt{(N)/t}}{N/t} = \frac{1}{\sqrt{(N)}} \qquad ...(25)$$

Account must be taken of the count rate due to background, unless the count rate of the sample is very large. Hence actual counting rate

$$RT = R_s + R_B,$$

where the subscripts refer to the total, sample and background respectively.

The standard deviation $\sigma s = \sqrt{(\sigma T^2 + \sigma n^2)}$

or $$\sigma s = \sqrt{\left(\frac{N_T}{tT^2} + \frac{N_B}{t_B{}^2}\right)} = \sqrt{\left(\frac{R_T}{tT} + \frac{R_B}{t_B}\right)} \qquad ...(26)$$

and $$V_S = \frac{\sigma s}{Rs} = \frac{\sigma s}{R_T - R_B} \quad ...(27)$$

SCINTILLATION COUNTERS

One of the earliest methods of detecting nuclear radiations was by the luminescence, they produced in certain substances. In 1903, Crookes in England and Elster and Geitel in Germany independently reported that alpha particles impinging on zinc sulphide screens produced individual flashes which could be observed through a microscope. The first attempt to count α-particles, by observing flashes of light (scintillations) generated in diamond by them, was made by E. Regener in Germany in 1908. In the same year *Rutherford and Geiger compared the number of scintillations seen with the number of pulses recorded by an ionization chamber. The numbers were same in both cases. Hence, if each particle caused a single pulse in the counter then it also gave rise to one scintillation.* In this way this method was available for counting individual α-particle. The observations of scintillations through low power microscope were very tedious and limited to relatively low counting rates. For this reason early forms of scintillation method soon became obsolete and more attention was paid to the development of gas counters.

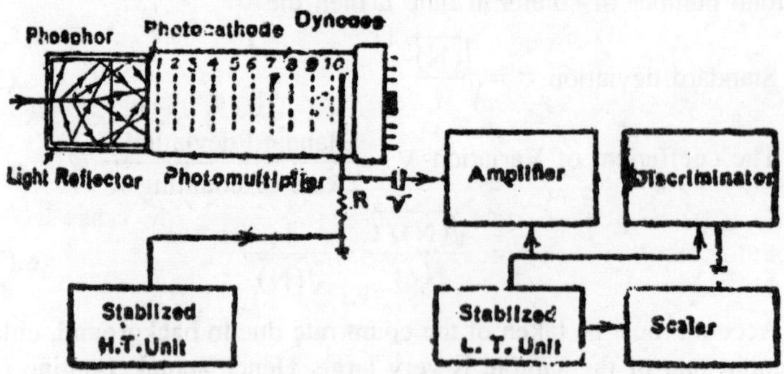

Fig. 19 : Schematic diagram of Scintillation Counter.

Since 1945, this old technique, is now playing a most important role. This has been due to the development of the photo-multiplier tube and the intensive study of the luminescent properties of many inorganic solids as well as organic compounds either in the solid state or in solution. A purely schematic diagram of a scintillation counter unit is shown in the following figure. A scintillation counter consists of the luminescent material (known is *scintillator*), reflecting layer such as aluminium foil enclosing the luminescent substance to facilitate the collection of light, light pipes, the photomultiplier tube, amplifier, voltage discriminator and an electrical circuit to record the output pulses. Before considering the performance of the scintillation counter let us first examine the behaviour of some of its components.

(a) Organic Liquids

Some solutions of organic compounds are also capable of acting as phosphors, although, they are not as efficient as the best pure crystals. The liquid scintillator has two main components: the solvent, (*e.g.*, toluene or xylene) and the primary solute (*e.g.*, a few percent of diphenyloxazole or terphenyl). Most of the nuclear radiation energy is absorbed by the solvent which is not a scintillator and is then transferred to the solute which emits the light actually. When the scintillations occur in the extreme ultraviolet region one uses a light shifter to increase the wavelength of the emitted light.

(b) Organic Crystal

Kallmann employed a large clear crystal of the organic substance naphthalene as phosphor. Following this, it has been found that other related organic compounds, consisting of several linked benzene rings are better scintillators. Of the solid organic scintillators anthracene and stilbene appear to be the best. The organic crystalline scintillators have a faster decay time ($\sim 10^{-8}$ sec), high transparency and poorer conversion efficiency especially for heavy particles. For β-particles response is linear. Because of low Z and ρ, these crystals have low cross-sections for photoelectric and pair production processes, hence can not be used for the detection γ-rays and X-rays.

(c) Inorganic Crystals

Unlike the organic compounds, inorganic substances do not scintillate when they are pure. Best one, especially for gamma rays, is crystalline *sodium iodide activated with thallium* (about 0.1 per cent). A less sensitive alternative is the nonhygroscopic cesium iodide, also activated with thallium. Calcium iodide with a trace of europium is more efficient than NaI, although, it is also hygroscopic. Silver activated zinc sulphide is an excellent scintillator but can be used only in thin layers, because it soon becomes opaque to the light emitted by it.

(d) Plastics

Organic scintillator system, which lies between a solid and liquid has a plastic as the base. A solution is made of the primary solute in a solvent like vinyl toluene or styrene which can be readily converted into a solid plastic. The resulting transparent material can be made in large pieces and cut to any desired shape. Plastic scintillators are widely used in high energy physics.

In a crystal such as NaI (T*l*) an ion Tl^+ replaces an ion Na^+ and forms an emission center. The T*l* ion thus occupies a position of equilibrium in the lattice such that a displacement from that position produces an increase in potential energy for the entire lattice system near the Tl^+ ion. The absorption of radiation changes the Tl^+ ion to an excited state for which the potential energy curve has a minimum at a position different from that in the ground state. The absorption (transition A) leaves the system in a state of high vibrational energy. The excited particle makes many collisions and with in 10^{-5} to 10^{-8} sec dissipates its surplus vibrational energy to the lattice. The most intense emission thus corresponds to transition A'. If the temperature is high enough, so that the higher vibrational levels of the ground state have appreciable population. Now, it is possible that the absorption takes place along B which is followed by emission of radiation.

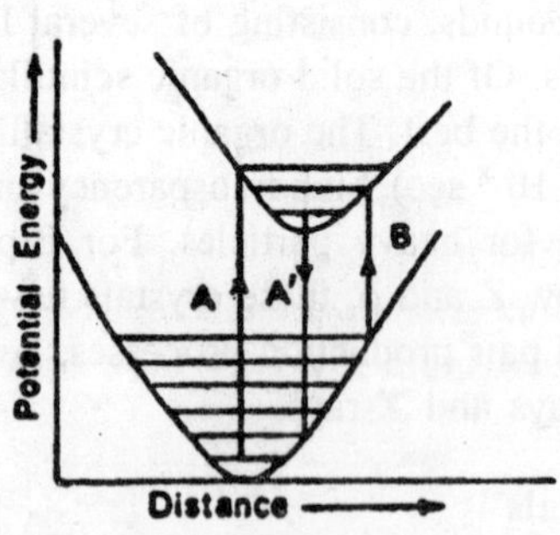

Fig. 20 : Luminescence proccess in NaI (Tl).

ZnS (activated with Cu or Ag)

1. The output is proportional to the energy of the incident particle.
2. C_{IP} is independent of the incident energy.
3. It has highest light conversion efficiency C_{IP} of any known phosphor.

4. The efficiency and emission spectra depend on the preparation and activator used.

For ZnS (activated with Cu)—conversion efficiency $C_{IP} = 28\%$ and the emission spectra from 4000-6000Å has a peak at 4500Å.

For ZnS (activated with Cu)— $C_{IP} = 25\%$ and $\lambda_{peak} = 5200$Å.

It has following limitations

1. Decay time is ~10^{-6} sec or longer.
2. A large crystal can not be prepared.
3. It has low transparency. Beyond thickness 25 mg/cm^2, the detector efficiency decreases.
4. It can not be used for the detection or γ-rays and electrons. Only the detection of slightly heavier particles such as a, p, d is possible.

NaI (activated with Tl)

1. Its absorption spectra lies in ultraviolet region, with the peaks at 2340 and 2930Å. Due to this reason, the crystals are highly transparent.
2. Its emission spectra has a average value of wavelength ~4100Å, and FWHM = 850 ± 100Å.
3. Its can be grown into large single crystal.
4. It has apart from ZnS, the highest.
5. Because of high values of p and Z, it has large cross sections for pair-production, photoelectric effect and compton effect. Due to this reason it is very suitable for the detection of gamma and X-rays.
6. The decay time at room temperature is lowest (~2.5×10^{-7}sec.).

C_{IP} decreases with increasing specific ionization causing a nonlinear response to α-particles. The response is linear to p, d and e. The crystal must be protected from moisture with surface film of liquid paraffin or be placed in dry atmosphere.

(e) Gases

Inert gases of the atmosphere, *e.g.*, argon, krypton and xenon when exposed to nuclear radiations produce scintillations in the ultraviolet region of the spectrum. To increase photomultiplier response the flashes are converted into visible light by means of a suitable wavelength shifter. An important advantage of the inert element scintillators iff that the light output is proportional very closely to the energy of the nuclear radiations.

Choice of Phosphors

NaI (Tl) is a highly effective scintillator as far as light output is concerned, but the light pulse has a relatively high decay time, compared with the organic phosphors. Hence *NaI (Tl) is less satisfactory when a very rapid response is required.* Iodine present in it gives a much greater stopping power for γ-rays. The size of the output pulse is proportional to the energy.

Organic scintillators give comparatively smaller light pulses, but have a shorter decay time. Hence, a higher counting rates are possible. The decay time in the liquid and plastic scintillators are even shorter than in the crystals Because of the small stopping power, the organic scintillators cannot be used for γ-rays unless the counter is large and are used for β-detection.

Screens with activated ZnS, in very thin layers, are useful for the detection of α-particles. Neutrons can be detected through the emission of a suitable charged particle resulting from a nuclear reaction initiated by the incident neutron.

Different Processes in the Scintillation Detector

(a) Absorption Process

If the dimensions of the scintillator are large compared to the range of the incident particle, the incident charged particle can dissipate all its energy to the scintillator. If the γ-ray is incident on a scintillator it may interact with the scintillator material in three ways: *photo-electric absorption, Comption scattering and pair production* each cases, electrons are produced. The all or part of the energy of incident γ-ray is transformed into the kinetic energy of these electrons.

(b) Scintillation Process

The electrons emitted will give up their kinetic energy in ionization or excitation in the scintillator material. This absorbed energy appears either as heat energy or as luminescence photons. In the latter cases the scintillator material de-excites by light emission within 10^{-8} seconds.

This emitted light is known as scintillation. If τ is the mean life of the scintillator, the number of light photons in time t after the ionization radiation has arrived is given by

$$n = \text{const}\ (1 - e^{-t/\pi}) \qquad ...(28)$$

(c) Formation of Electrical Pulses

The light produced in scintillator falls on the photocathode of the photomultiplier tube, producing photoelectrons. These electrons fall on the first dynode, producing a bunch of secondary electrons. This process is repeated and a gain upto $10^{-7} - 10^{8}$ is achieved. The electrons finally fall on the electrode, known as anode, and thus, produce an electrical pulse across the load resistance R. The pulse height V at the output related to the amount of charge Q collected at the output through the relation

$$V = \frac{Q}{C} \qquad ...(29)$$

where C is the capacity of the output point.

If E_i be the energy of the incident charged particle or γ-rays and A the probability of amount of energy absorbed by the phosphor.

∴ Energy dissipated in the phosphor = E_iA.

This energy is converted within efficiency C_{IP} into photons of average energy E_p.

$$nr = E_iA\ C_{IP}/E_p \qquad ...(30)$$

If T_p is the optical transparency and G is the fraction of the photons falling on the photocathode. Hence, the number of photons incident on the photocathode.

$$n_p/ = T_pGn_p = E_iAC_{IP}T_pG/E_p.$$

These photons are converted into photoelectrons with an efficiency C_{PE} f(v), where C_{PE} is the photoelectric conversion efficiency of the cathode and f(v) is the relative response at frequency v.

∴ No. of photo electrons n = $C_{PE}/(v)$ np/

If f_d be the fraction of photoelectrons reaching the first dynode and M the electron multiplication factor, then the total number of electrons reaching the anode

$$N = f_d Mn = C_{PE} f(v)\, E_i A C_{IP} T_p G f_d M/E_p.$$

Hence total charge Q = eN and the pulse height

$$V = Q/C = eE_i\, AC_{IP} T_P G C_{PE} f(v) f_d M/CE_p \qquad ...(31)$$

Here, we see that the pulse height is proportional to the incident particle if the particle is absorbed in the scintillator completely.

Resolving Power

The height of the output pulse exhibits variations about the mean value for a given particle energy E. This variation is due to the statistical fluctuations in the various factors of eqn. (31). The resolution R of a scintillation counter can be defined as follows

$$\frac{1}{R} = \frac{\overline{Q}^2 - (\overline{Q}^2)}{(\overline{Q}^2)} \qquad ...(32)$$

where $\overline{Q}$ is the mean charge collected at the anode and $\overline{Q}^2$ is the mean squared charge collected at the anode. The smaller the half width of the photo peak, the higher will be energy resolution R.

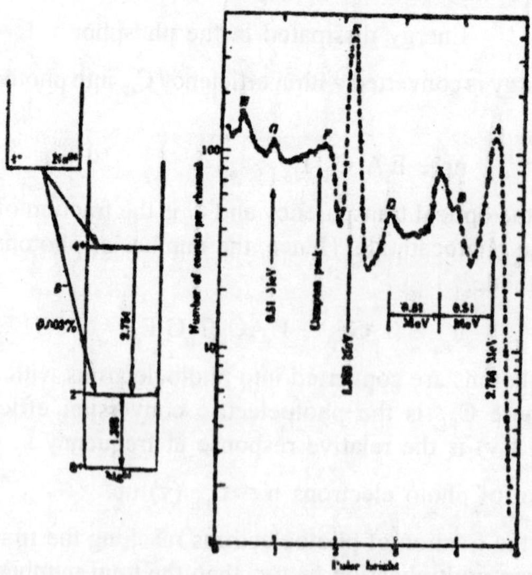

Fig. 21 : Decay scheme for Na^{24} (Left). Pulse height spectrum for gamma rays (Right).

The scintillation counter can be used as a spectrometer in conjunction with an electronic pulse height analyzer, which has found many applications in low energy, as well as in high-energy nuclear physics. Suitable circuits have been designed to sort out particles with in a given range of energy so that distribution in energy of the incident radiation may be measured. Unfortunately, even mono-energetic γ-rays produce a complex pulse height spectrum. The pulse height spectrum produced by γ-rays from the lower excited states of Mg^{24} is shown in Fig. 22 (right). This isotope is produced in low lying states by β-emission of Na^{24} according to the decay scheme shown in Fig. 27 (left). Two γ-rays of energy, one at 2.754 MeV and other at 1.368 MeV are emitted. The peak A in this spectrum corresponds to the full energy of 2.754 γ-ray and is due to pair production process only.

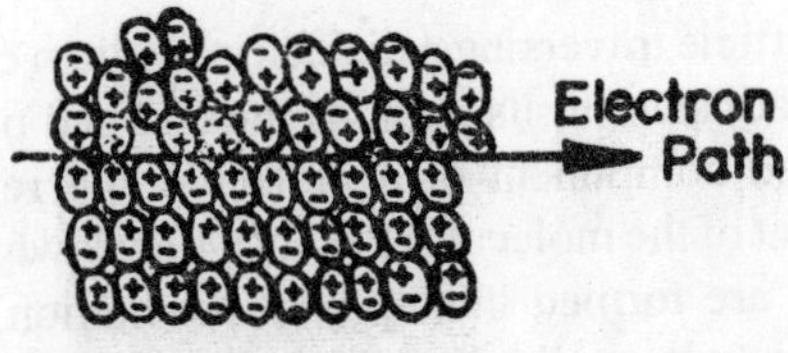

Fig. 22 : Polarization of molecules.

The next peak B corresponds to the electrons produced by compton process. The peak C is caused by pair-production processes in which one of the two annihiliation photons has escaped and peak D corresponds to the processes in which both the annihiliation photons have escaped.

The central sharp peak E corresponds to the energy 1.368 MeV and is due to pair production or photo electric process. The compton peak corresponding to this peak is F. As 1.368 MeV pair production is relatively unimportant, hence first and second escape peaks corresponding to the energies 1.368 – 0.51 = .858 and 1.368 – 1.02 = .348 are not observed. Peak G corresponds to 0.51 MeV annihiliation photons produced outside the scintillator but absorbed by it. The peak H arises from back scattered photons from compton events.

The scintillation counter has a much higher counting efficiency for gamma rays due to the greater amount of energy dissipation by the gamma rays. The scintillation counter offers greater stability, greater accuracy, shorter resolving time and higher efficiency than G.M. counter. Provided all the particle energy is dissipated in the scintillator, the voltage pulse height produced can be used to measure the particle energy, once the counter is calibrated by means of a standard source. Because of their extremely rapid response scintillation counters are having many applications in atomic energy field. These are used for the accurate timing of nuclear particles moving with very high speeds.

CERENKOV COUNTERS

When a charged particle moves through a transparent dielectric medium with a velocity greater than the velocity of light in that medium, weak electro-magnetic radiation is emitted. These radiations are named Cerenkov radiations. The radiation is emitted mainly in the direction of the charged particle and has a continuous spectrum from red to ultra-violet at least.

A charged particle traversing a dielectric medium causes polarization of the molecules adjacent to its path, as the charged particle attracts the oppositely charged component of a molecule and repels the similarly charged component of the molecule (Fig. 28). For slowly moving particles, polarized regions are formed in step with the motion in the vicinity of the particle. The resultant dipole field is zero as the polarization field around the electron is symmetrical in the axial and azimuthal directions, hence no radiation is emitted. For fast moving particles the polarization is *axially asymmetric* but azimuthally symmetric and the resultant dipole field acts even at large distances from the electron's track. The radiation pulses are thus emitted in succession by the molecules along the path of the charged particle. These radiations are like the radiations emitted from a *damped oscillating dipole moment.*

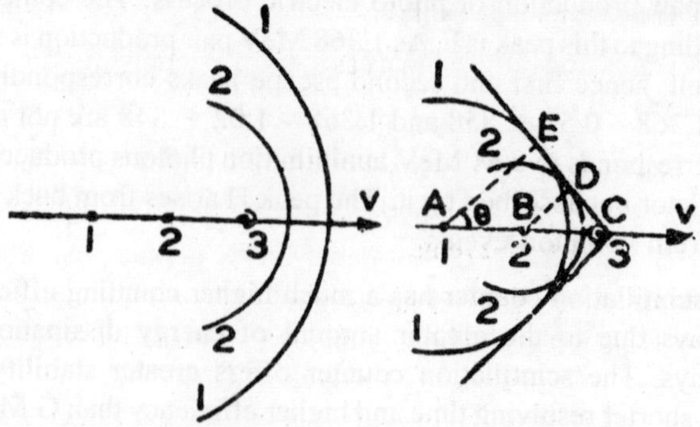

Fig. 23 : Wavefronts resulting from radiation losses: with v < c/μ (left) and with v > c/μ (right).

In a dielectric medium of refractive index μ, photons move with a velocity c/μ. Fig. 23 (left) shows the three interactions between a charged particle moving with velocity v (< c/μ) and the electrons in the medium. The wavefronts of the electromagnetic waves resulting from interaction do not interrelate and these interactions are entirely independent. Fig. 23 (right) shows the interactions when v > c/μ. The wavefronts from 1 and 2 lie inside 3 at the time of the third interaction.

The Huygens construction of elementary wave optics shows that a conical wavefront can be drawn tangent to the spherical surfaces and the Cerenkov radiation can be observed at a particular angle with respect to the direction of motion of the particle. The particle moves along ABC with velocity v It radiates energy in all directions from the points along its path ABC. The time in which particle travels a distance AC, the radiations will move a distance AE, CE is thus the wavefront. The angle which the Cerenkov Radiation makes with the direction of charged particle is given by

$$\cos\theta = \frac{AB}{AC} = \frac{BD}{BC} = \frac{ct/\mu}{vt} = \frac{c}{\mu v} = \frac{1}{\beta\mu} \qquad \text{...(33)}$$

The Cerenkov radiation is not confined to one plane but is propagated along the surface of a cone whose axis coincides with the direction of motion of the charged particle and whose semivertical angle is θ, given by eqn. (33), which shows that:

1. θ depends on the frequency (*i.e.,* wavelength) of the Cerenkov radiation. The radiation arc mainly of the high frequency region of the visible spectrum.
2. Unlike the bremsstrahlung, θ is independent of the mass of moving charged particle.
3. The emission angle increases with increasing the particle velocity β.
4. As P cannot exceed unity the maximum angle of emission.

   $$\theta_{max} = \cos^{-1}(1/\mu)$$
5. For a given constant refractive index μ, there is a threshold velocity β_{th} = l/μ below which no Cerenkov radiation is emitted and at which the radiations are along the particles direction of motion.

Radiation (33) for the constructive interference should satisfy the following conditions:

(a) The particle track length must be large compared with the wavelength λ of the Cerenkov radiation.

(b) The particle must move with the constant velocity in the medium.

Frank and Tamm have shown from classical theory that the total energy radiated in a short element of the particle's path ds is given as

$$\frac{dT}{ds} = \frac{\pi z^2 e^2}{\epsilon_o c^2} \int \left(1 - \frac{1}{\beta^2 \mu^2}\right) v\, dv$$

where ze is the charge on the moving charged particles, v is the frequency of the emitted radiation, and the integral extends overall frequencies for which $\beta\mu > 1$. This loss is quite small (*e.g.*, about 10^3eV/cm for singly charged particles moving through glass or lucite) and is negligible compared to the ionization losses.

Consider the Cerenkov radiation emitted between two frequencies v_1 and v_2 as composed of quanta whose average energy is $hv = 1/2h\,(v_1 + v_2)$, then from eqn. (44) the average number of quanta emitted per unit length is

$$\frac{1}{h_v}\frac{dT}{ds} = \frac{\pi z^2 e^2}{\epsilon_o hc^2}(v_2 - v_1)\left(1 - \frac{1}{\beta^2 \mu^2}\right) \quad ...(35)$$

where μ if the average refractive index of the medium over the frequency interval v_1 to v_2.

Relation (35) shows that for extreme relativistic electrons passing through glass ($Z = -1$, $\beta \simeq 1$, $\mu \simeq 1.5$) the (Cerenkov angle is 48° and the Cerenkov photons emitted per cm within the visible spectrum ($\Delta v = v_2 - v_1 \simeq 3 \times 10^{14}\ sec^{-1}$) are 200. As the number of quanta per unit path length varies as I/λ^2, the short wavelength are thus predominantly present in Cerenkov radiation.

The properties discussed above have been extensively utilized in Cerenkov detectors which can be used not only to detect swiftly moving charged particles but also to determine their velocity. These detectors possess the following advantages:

1. Dependence of intensity on particle velocity.
2. Directional emission of light, with an angle dependent on velocity.
3. High efficiency and high counting rate.

4. The duration of the light pulse at any point with in the Cerenkov cone is vanishingly short ($\sim \leq 50^{-10}$ sec).

R.L. Mather has observed Cerenkov radiations radiated by a collimated beam of 340 MeV protons passing through an optically flat sheet of dense flint glass of 0.67 mm thickness with a 35 mm Leica camera. The glass sheet, whose rear surface was alumunized, was set with respect to the proton beam so that part of the Cerenkov cone radiated in the glass would incident normally on the aluminium surface and reflect in such a way that there would be no refraction at the front surface of the glass. The angle of emission of radiations varies with wavelength λ because the refractive index depends on λ. This dispersion effect was decreased to nearly zero by Mather after introducing a prism. The recorded Cerenkov radiation on the camera film corresponded to a θ value of 38.5°. Using the relation (33) β comes out to be 0.680. The kinetic energy E_{kin} of the protons can be obtained as

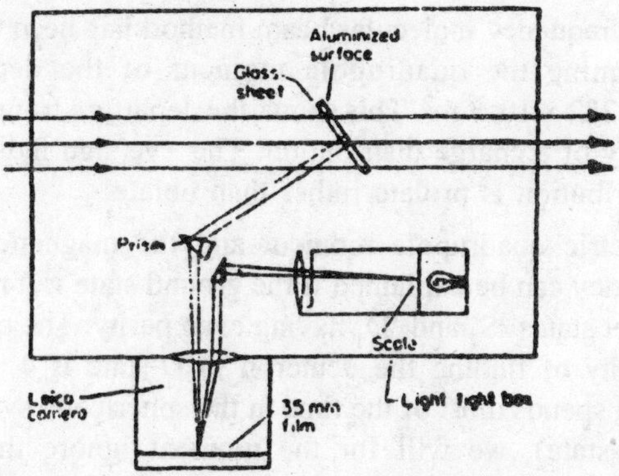

Fig. 24 : Measurement of velocity of high energy protons.

$$E_{kin}\ m_p c^2 (1 - \beta^2)^{-1} - m_p c^2 + 341 \text{ MeV} \qquad ...(36)$$

This value was found to have 1% accuracy.

DEUTERON

The deuteron does possess measurable properties which might serve as a guide in the search for the correct nuclear interaction. These properties are:

1. The parity of deuteron as measured, indirectly, by studies of nuclear disintegrations and reactions for which certain rules of parity changes exist, is even.

2. The angular momentum quantum number, often called the nuclear spin, of the ground state of the deuteron determined by a number of optical, radio frequency and micro-wave methods is one. It suggests that the spins are parallel (triplet state) and the orbital angular momentum of the deuteron about their common center of mass is zero. Thus, the ground state is $^3 5$ state.
3. The binding energy of the deuteron is very small. Its experimental value is 2.225 ± 0.002 MeV. Since the energy needed to pull a nucleon out of a medium mass nucleus is about 8 MeV, we must regard the deuteron as loosely bound.
4. The extraordinary stability of the alpha particle shows that the most stable nuclei are those in which number of neutrons and photons are equal. The deuteron consists of two particles of roughly equal masses M, so that the reduced mass of the system is 1/2 M.
5. A radio frequency molecular beam method has been employed to determine the quadrupole moment of the deuteron as $Q = 0.00282 \times 10^{-28}\ m^2$. This shows the departure from spherical symmetry of a charge distribution. The +ve sign indicates that this distribution is prolate rather than oblate.

   The electric quadrupole moment and the magnetic moment discrepancy can be explained if the ground state is a mixture of the triplet states $^3S_1$ and $^3D_1$ having even parity. The percentage probability of finding the deuteron in D-state is 4 ± 2%. As deuteron spends most of the time in the spherically symmetrical state (S-state), we will for the moment ignore the D-state contribution to the deuteron wave function.
6. The sum of the magnetic dipole moments of the proton (2.79275 μN) and neutron (–1.91315 μN), do not exactly equal to magnetic moment of the deuteron (0.85735 μN) measured by magnetic resonance absorption method.
7. Since the neutron has no charge, the force between the neutron and proton can not be electrical. This force can not be magnetic as magnetic moments are very small. It can not be gravitational force, as the masses are very small. So we must accept the nuclear force as a new type of force. This force is short range, attractive and along the line joining the two particles (central force). Since a central force can not account for the quadrupole

moment of the deuteron. As a quadrupole moment is small, the assumption will be approximately correct.

8. The force depends only on the separation of the nucleons not on the relative velocity or orientation of the nucleon spins with respect to the line. This force can be derived from a potential. Since the force is attractive, V(r) is negative and decreases with decreasing r. Since it is short range, V(r) vanishes for r > b, where b ~ 3 fermi.

The Schrodinger wave equation for the two body problem is

$$\Delta^2\phi + (2m/h^2)\,(E - V)\psi = 0, \qquad ...(1)$$

where m is the reduced mass, E the total energy of the system equal to the binding energy of deuteron and V the potential energy describing the forces acting between the two bodies.

In this terms of spherical polar coordinates, eqn. (1) becomes

$$\left[\frac{1}{r^2}\frac{\partial}{\partial r}\left(r^2\frac{\partial\phi}{\partial r}\right) + \frac{1}{r^2\sin\theta}\frac{\partial\theta}{\partial}\left(\sin\theta\frac{\partial\phi}{\partial\theta}\right) + \frac{1}{r^2\sin^2\theta}\frac{\partial^2\psi}{\partial^2\phi}\right]$$

$$+ (2m/\hbar^2)\,[E - V(r, \theta, \phi)]\,\psi = 0 \qquad ...(2)$$

Let V (r, θ, ϕ) actually depends on r only and not on θ and ϕ. The solution of the above equation can be written as a product of a function of r only and one of θ and ϕ only as ψ(r, θ, ϕ) = ϕ (r) ϕ (θ, ϕ). Substituting it into equation (2) we get

$$\frac{1}{\psi(r)}\frac{d}{dr}\left(r^2\frac{d\phi(r)}{dr}\right) + \frac{2mr^2}{\hbar^2}[E - V(r)] = -\frac{1}{\psi(\theta, \phi)}$$

$$\left[\frac{1}{\sin\theta}\frac{\partial}{\partial\theta}\left(\sin\theta\frac{\partial\phi(\theta,\phi)}{\partial\theta}\right) + \frac{1}{\sin^2\theta}\frac{\partial^2\psi(\theta,\phi)}{\partial\phi^2}\right] \qquad ...(3)$$

The L.H.S. of this equation depends only on r, the R.H.S. depends only on θ and ϕ. For all values of the variables, each side of this must separately equal to the some constant, which comes out to be $l(l + 1)$. Thus we get

$$\frac{1}{r^2}\frac{d}{dr}\left(r^2\frac{d\psi(r)}{dr}\right) + \frac{2m}{\hbar^2}\left[E - V(r) - \frac{l(l+1)\,\hbar^2}{2mr^2}\right]\psi(r) = 0$$

$$...(4)$$

where *l* is the angular momentum quantum number of the system.

The last term in the bracket appears as a straight addition to the actual potential V(r) and is known as centrifugal potential.

The Schrodinger equation for ^3S state ($l = 0$) of the deuteron is

$$\frac{1}{r^2}\frac{d}{dr}\left(r^2\frac{d\psi(r)}{dr}\right)+\frac{2m}{\hbar^2}[E-V(r)]\,\psi(r)=0 \qquad ...(5)$$

In this case reduced mass m = 1/2 M. We expect the ground state to be spherically symmetric (S-state), so that $\phi(r)$ depends only or r. Substituting $\psi(r) = u(r)/r$ in equation (5), where u(r) is called the radial wavefunction, we get

$$\left(\frac{d^2u}{dr^2}\right)+\left(\frac{M}{\hbar^2}\right)[E-V(r)]\,u=0 \qquad ...(6)$$

The wavefunction of the bound state of the deuteron is not markedly dependent on the exact shape of the potential V(r) between a proton and neutron provided that a potential of short range is chosen. For simplicity, we represent V(r) by a *square well* of depth V_0 and radius b, where b is the range of the nuclear force. In it V(r) has a constant negative value $-V_0$ for separations less than a certain value b and the value zero for all greater separations. The other central force potentials may be of

Gaussian well type $\qquad V(r) = -V_0e^{-(r/\alpha)^2}$

Exponential well type $V(r) = -V_0e^{-r/\alpha}$

Yukawa well type $\qquad V(r) = -\dfrac{V_0e^{-r/\alpha}}{(r/\alpha)}$

For the ground state of the deuteron, the total energy E is negative and equal to – B, where B is the binding energy of deuteron. Thus, equation (6) can be written as

$$\frac{d^2u}{dr^2}+\frac{M}{\hbar^2}[V_0-B]u=0 \text{ for } r<b \qquad ...(7)$$

and

$$\frac{d^2u}{dr^2}+\frac{M}{\hbar^2}(-B)u=0 \text{ for } r>b \qquad ...(8)$$

These equations can be written as

$$\frac{d^2u}{dr^2} + K^2u = 0, \quad r < b \qquad ...(9)$$

and
$$\frac{d^2u}{dr^2} - \alpha^2u = 0, \quad r > b \qquad ...(10)$$

where
$$K^2 = \frac{M(V_0 - B)}{\hbar^2}$$

and
$$\alpha^2 = \frac{MB}{\hbar^2}$$

General solutions of eqns. (9) and (10) are given as

$$u = A_1 \sin Kr + B_1 \cos Kr \qquad ...(11)$$

$$u = A_2e^{\alpha r} + B_2e^{-\alpha r} \qquad ...(12)$$

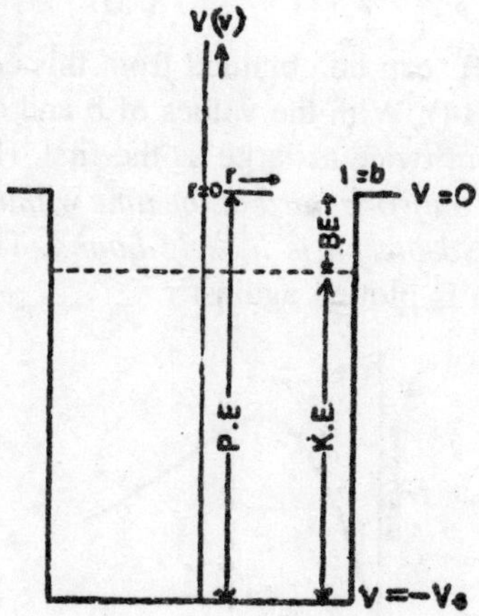

Fig. 26 : Square well potential of deuteron.

The following boundary conditions must be imposed:

1. u (r → 0) = 0, to keep wave function y finite.
2. u (r → ∞) = 0, u must not diverge faster than r as r → ∞.

To satisfy the conditions at zero and infinity, the solutions reduce to

$$u = A_1 \sin Kr \qquad \text{for } r < b \ ...(13)$$

$$u = B_2e^{-\alpha r} \qquad \text{for } r > b \ ...(14)$$

Since these two solutions join smoothly at r = b. Hence equating the values and first derivatives of u at r = b, we have

$$A_1 \sin Kb = B_2 e^{-\alpha b} \quad ...(15)$$

and $A_1 K \cos Kb = - B_2 e^{-\alpha b}$ ...(16)

$$\therefore \quad K \cot Kb = -\alpha \quad ...(17)$$

The constants A_1 and B_2 are obtained from the requirement that the integral of $|\psi|^2$ over all space must be equal to unity.

$$4\pi \int_0^{\infty} |\psi|^2 r^2 dr = 4\pi \int_0^{\infty} u^2 dr = 1$$

or $$4\pi \int_0^b A_1^2 \sin^2 Kr dr + 4\pi \int_0^{\infty} B_2^2 e^{-2\alpha r} dr = 1$$

$$A_1^2 [b - (1/2K) \sin 2Kb] + \left(\frac{B_2^2}{\alpha}\right) e^{-2\alpha b} = \frac{1}{2}\pi \quad ...(18)$$

Thus, A_1 and B_2 can be obtained from this equation with the help of eqns. (13) and (14). With the values of b and α as given above, the second term is about twice as large as the first. Hence the *nucleons in the deuteron spend only one third of the time within the range of nuclear force and thus the deuteron is loosely bound.* This can be seen from Fig. 27, where u(r) is plotted against r.

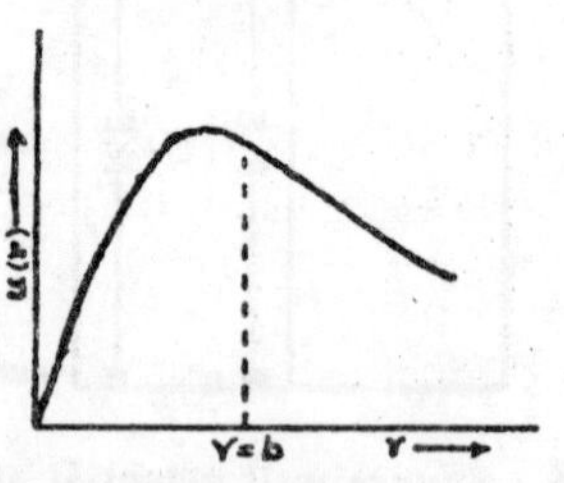

Fig. 27 : Ground state deuteron wave function.

SOLVED EXAMPLES

Example 1:

A Geiger-Müller counter has a plateau slope of 3% per 100 volts. If the operative point is at 1100 volts, what is the maximum permissible voltage fluctuation if the counting is not to be affected by more than 0.1?

Solution:

As the plateau slope $= \frac{n_2 - n_1}{n_{av}} \times \frac{100}{V_1 - V_1} \times 100\%$ per 100 volts

In this problem $(n_2 - n_1)/n_{av} = 0.1\% = 0.1 \times 10^{-2}$ and slope = 3

$\therefore V_2 - V_1 = 0.1 \times 10^{-2} \times 1100 \times 100/3 =$ **3.3 volts.**

As operating voltage is 1100 volts, hence the voltage fluctuation is 3.3 volts in 1100 volts or 0.3%.

Example 2:

An organic-quenched G.M. tube operates at 1000 volts and has a wire diameter of 0.2 mm. The radius of the cathode is 2 cm, and the tube has a guaranteed life-time of 10^9 counts. What is the maximum radial field and how long will the counter last if it is used on the average, for 60 hours per week at 2000 counts per minute?

Solution:

The maximum radial field or the field along a radius at the wire surface

$$E_{maz} = \frac{V_0}{2.3026 \times \log_{10}(b/a)} = \frac{1000}{2.3026 \times 10^{-4} \times \log_{10} 200}$$

$$= 1.8^9 \times 10^6 \text{ Volts/metre.}$$

If the lifetime of the tube is N years the total number of counts recorded will be equal to the guaranteed counts. Hence

$$N \times 50 \times 60 \times 60 \times 2000 = 10^9$$

or **N = 2.77 years.**

Example 3:

A.G.M. tube with a cathode 5.0 cm. in diameter and a wire diameter of 0.012 cm is filled with argon to a pressure such that the mean free path is 7.8×10^{-4} cm. Calculate the value of the voltage that must be applied to just produce an Avalanche.

Solution:

Electric field strength E at any radius r between two concentric cylinders, when a potential V is applied is given by

$$E = \frac{V}{r \log_e (b/a)}$$

Avalanche will just, start when the integral of this equation, taken over one mean free path is equal to the ionisation potential of the filling gas, *i.e.*,

$$E_{a\lambda} = \frac{V\lambda}{a \log_e (b/a)} = I \quad \text{or} \quad \frac{Ia \log_e (b/a)}{\lambda}$$

$$V = \frac{15.7 \times 0.006 \times 10^{-2} \times 2.3026 \log_{10} (0.025/0.00006)}{7.8 \times 10^{-6}}$$

$$= \mathbf{730\ volts.}$$

EXERCISES

1. A sample of uranium, emitting a-particles of energy 4.18 MeV, is placed near an ionization chamber. Assuming that only 10 particles per sec. enter the chamber, calculate the current produced.
2. The output circuit of a ratemeter consists of a resistance of 50,000 ohms shunted by a condenser. The output pulses from the ratemeter have a constant amplitude of 5.0 milliamperes, and a variable width. What pulse width is required to produce a voltage of 5.0 volts across the tank circuit at a count rate of 200 cpm? What pulse length will be required to obtain the same voltage at a count rate of 50,000 cpm? What capacitance will be required to obtain a circuit time constant of 2 sec ?
3. A square wave whose peak to peak value is 1.0 volt extends 0.5 V with respect to earth. The duration of the positive section is 0.1 sec and of the negative section is 0.2 sec. If this waveform is impressed upon a R-C differentiating network of time constant 0.2 sec, what are the steady state maximum and minimum values of the output waveform.